Herpes Zoster

Monographs in Virology

Vol. 26

Series Editor

H.W. Doerr *Frankfurt a.M.*

From the Herpes Management Forum of the Paul-Ehrlich-Gesellschaft

KARGER

Herpes Zoster

Recent Aspects of Diagnosis and Control

Volume Editors

Gerd Gross *Rostock*
Hans-Wilhelm Doerr *Frankfurt a.M.*

52 figures, 24 in color, and 18 tables, 2006

KARGER

Basel · Freiburg · Paris · London · New York ·
Bangalore · Bangkok · Singapore · Tokyo · Sydney

Prof. Gerd Gross
Department of Dermatology and
Venerology, Director
University of Rostock
Augustenstrasse 80–84
DE–18055 Rostock (Germany)

Prof. H.W. Doerr
Institute of Virology, Director
University of Frankfurt
Paul-Ehrlich-Strasse 40
DE–60596 Frankfurt am Main (Germany)

Library of Congress Cataloging-in-Publication Data

Herpes zoster : recent aspects of diagnosis and control / volume
editors, Gerd Gross, Hans-Wilhelm Doerr.
p. ; cm. – (Monographs in virology ; v. 26)
Includes bibliographical references and index.
ISBN-13: 978-3-8055-7982-7 (hard cover : alk. paper)
ISBN-10: 3-8055-7982-9 (hard cover : alk. paper)
1. Shingles (Disease) 2. Ophthalmic zoster. 3. Varicella-zoster
virus. I. Gross, G. (Gerd) II. Doerr, Hans Wilhelm. III. Series.
[DNLM: 1. Herpes Zoster. 2. Herpes Zoster–complications.
3. Herpes Zoster–therapy. W1 MO569P v.26 2006 / WC 575 H563
2006]
RC147.H6H47 2006
616.9'112–dc22

2006025294

Bibliographic Indices. This publication is listed in bibliographic services, including Current Contents® and Index Medicus.

Disclaimer. The statements, options and data contained in this publication are solely those of the individual authors and contributors and not of the publisher and the editor(s). The appearance of advertisements in the book is not a warranty, endorsement, or approval of the products or services advertised or of their effectiveness, quality or safety. The publisher and the editor(s) disclaim responsibility for any injury to persons or property resulting from any ideas, methods, instructions or products referred to in the content or advertisements.

Drug Dosage. The authors and the publisher have exerted every effort to ensure that drug selection and dosage set forth in this text are in accord with current recommendations and practice at the time of publication. However, in view of ongoing research, changes in government regulations, and the constant flow of information relating to drug therapy and drug reactions, the reader is urged to check the package insert for each drug for any change in indications and dosage and for added warnings and precautions. This is particularly important when the recommended agent is a new and/or infrequently employed drug.

www.karger.com
Printed in Switzerland on acid-free paper by Reinhardt Druck, Basel
ISBN-10: 3–8055–7982–9
ISBN-13: 978–3–8055–7982–7

Contents

Preface

Herpes zoster is a serious neurocutaneous disease which has been underestimated in terms of its burden, particularly in the elderly population. The overall incidence of herpes zoster in Europe is about 3 per 1,000 people and more than 10 per 1,000 people per year aged more than 80 years. Due to the growing life expectancy of the Central European population, the incidence of herpes zoster and its burden is very likely to increase further in the near future.

The onset of herpes zoster is almost always associated with waning varicella–zoster virus (VZV) specific cellular immunity. Herpes zoster is characterised by a more or less painful vesicular rash usually restricted to a defined area of the skin (dermatoma), which is innervated by the branches of a single sensory or a cranial nerve. The most common debilitating complication of herpes zoster is postherpetic neuralgia, which is especially seen in patients beyond 50 years of age.

The large majority of zoster cases are seen in elderly people. Nevertheless, a significant and increasing number is also diagnosed in young adults, in children and even in infants. As a rule of thumb, the risk to fall ill with zoster can be estimated as one-fourth of the age life span in years. In very early records, an association was recognized between herpes zoster and varicella, which is an ubiquitous and easily transmissible disease. After varicella has been clinically distinguished from smallpox, Steiner [1] postulated an identical infectious agent causing both herpes zoster and varicella. Kundratitz [2] described an identical histology of vesicles in varicella and zoster. Smears of affected cells show intranuclear inclusions in contrast to smears of smallpox vesicles presenting cytoplasmatic inclusions. Ruska, who had invented electron microscopy, first described the ultrastructurally indistinguishable morphology of herpes zoster virus and varicella virus [3]. In the same year (1943), herpes zoster was supposed by Garland [4] to be due to the reactivated varicella virus infection. It took 10 more years, until Weller and Stoddard [5] succeeded in developing suitable cell cultures for the isolation of the herpes zoster agent. They and others detected indistinguishable cytopathological effects caused by this and by varicella virus. This finally leads to the description of VZV as identical virus causing both varicella and herpes zoster. For a long time herpes simplex and herpes zoster were considered to be different manifestations of basically the same recurrent infection. The term herpes describes microefflorescences on the cornea both of herpes simplex and herpes zoster patients (herpes reflects the Greek word herpein = to creek). The development of methods to propagate VZV in cell cultures helped to establish diagnostic serology and to determine the VZV-specific humoral immune status. These investigations revealed that preexisting VZV-serum antibodies protect against varicella, but not against herpes zoster. On the contrary, people without antibodies do not develop herpes zoster. With other words: Herpes zoster virus could only be isolated from people with a history of varicella confirmed by the detection of serum antibodies. Serologic assays also revealed the difference between VZV and HSV. Using molecular biological techniques (DNA—DNA hybridization) VZV was grouped into the same subfamily of alpha-herpesvirinae as HSV. Restriction endonuclease analysis of genomic DNA extracted from VZV strains isolated from patients, who first had varicella and later on herpes zoster, definitely proved the identity of the virus and its reactivation in the same patient [6].

Occurrence and course of herpes zoster and herpes simplex are strikingly different: While in predisposed individuals herpes simplex is frequently recurrent throughout life, herpes zoster is usually a unique disease of the elderly and of immunocompromised individuals of any age. Much more than herpes simplex, the eruption of herpes zoster is rather strictly correlated to a waning antiviral cell-mediated immunity. Thus, the diagnosis of zoster in a patient younger than 50 years demands to check for an immunocompromising disease such as leukaemia, Morbus Hodgkin, HIV-infection, AIDS etc. Herpes zoster in infants is a rather rare finding. Commonly it results from a prenatally or perinatally acquired VZV infection, when the cell-mediated immune system of the newborn is still immature.

Despite the advent of antiviral therapy, herpes zoster remains a challenge for both physicians and scientists. In particular in older people, the rate of severe herpes zoster complications is increasing, e.g. meningitis, less frequently encephalitis and optic nerve damage. Zoster may be associated with chronic pain, so called postherpetic neuralgia, which is especially harmful in the head region innervated by the trigeminus nerve. The ganglion Gasseri is a predilection site of VZV latency and similarly also of HSV latency. Nucleic acids of both herpes viruses have been detected simultaneously at this site. However, in contrast to HSV, VZV may be present also at all spinal ganglia after primary VZV-infection (chickenpox).

To fight the complications, it is mandatory to establish rapid clinical and, if necessary, laboratory diagnosis and to begin antiviral therapy in time. Correct diagnosis and indication of therapy challenge dermatologists, neurologists, ophthalmologists and otologists. This lead to controversial discussions in the past. Thus, medical and scientific societies in many countries have established specific guidelines [7, 8].

Actually scientific interest focuses on VZV persistence. Similar to other herpesviruses two different forms of persistence seem to exist: (a) Proviral latency, which means genomic persistence without virus production and (b) low level VZV production. In this context immune escape has to be elucidated. It is obvious, that investigations, which study how VZV genome transcription is switched on and off, have great pharmaceutical relevance. This is particularly true with regard to the development of vaccines and new antiviral therapies.

Based on the preparation of clinical and laboratory medical guidelines for the management of zoster patients, the editors of this book intended to bring together leading specialists of clinical and scientific disciplines in order to compile the various insights and experiences concerning herpes zoster and VZV. It seems to be very useful at this time to present the state of the art and to describe the direction of further research activities, which will be focused on very early prevention of chronic zoster pain by a combined antiviral and analgetic therapy and on prevention of herpes zoster by use of a VZV-specific zoster vaccine.

H.W. Doerr, Frankfurt am Main
G. Gross, Rostock

References

1 Steiner G: Zur Inokulation der Varicellen. Wien Med Wochenschr 1875;25:306.
2 Kundratitz K: Über die Ätiologie der Varicellen und ihre Beziehung zum Zoster. Wien Klin Wochenschr 1925;38:502–503.
3 Ruska H: Über das Virus der Varicellen und des Zoster. Wien Klin Wochenschr 1943;22:703–704.
4 Garland J: Varicella following exposure to herpes zoster. N Engl J Med 1943;228:336–337.
5 Weller TH, Stoddard MB: Intranuclear inclusion bodies in cultures of human tissue inoculated with varicella vesicle fluid. J Immunol 1952;68:311–319.
6 Straus SE, Reinhold W, Smith HA, Ruyechan WT, Henderson DK, Blaese RM, Hay J: Endonuclease analysis of viral DNA from varicella and subsequent zoster infection in the same patient. N Engl J Med 1984;311:1362–1364.
7 Gross G, Schöfer H, Wassilew S, Friese K, Timm A, Guthoff R, Pau HW, Malin JP, Wutzler P, Doerr HW: Herpes zoster guideline of the German Dermatology Society (DDG). J Clin Virol 2003;26:277–289.
8 Volpi A, Gross G, Hercogova J, Johnson RW: Current management of herpes zoster: The European view. Am J Clin Dermatol 2005;6:317–325.

Gross G , Doerr HW (eds): Herpes Zoster.
Monogr Virol. Basel, Karger, 2006, vol 26, pp 1–8

Molecular Biology of Varicella–Zoster Virus

Markus Rahaus, Nathalie Desloges, Manfred H. Wolff

Institute of Microbiology and Virology, Private University of Witten-Herdecke, Witten, Germany

Morphology of the Virion and Genome Organisation

Varicella–zoster virus (VZV), also known as human herpesvirus 3 (HHV3) belongs to the herpesvirus family (Herpesviridae). This classification is based on the morphological characteristics of the virus and its physical and chemical properties. The Herpesvirus Study Group of the International Committee on the Taxonomy of Viruses (ICTV) divided the members of this family into three subfamilies: Alphaherpesvirinae, Betaherpesvirinea and Gammaherpesvirinea. Based on its host spectrum, the length of the replicative cycle, the cytopathic effect in vitro and the particularities in the establishment of latency, VZV together with herpes simplex virus type 1 (HSV1; HHV1) and type 2 (HSV2; HHV2) were grouped into the subfamily of Alphaherpesvirinae. Moreover, by its genome organisation, VZV was classified into the genus varicellovirus, whereas HSV was classified into the genus simplexvirus, for overview see [1].

Though symptoms of an infection with one of these herpesviruses differ strongly from each other, the morphology of the particles and the biological properties are very similar. VZV is characterised by a strongly limited spectrum of infectable host cells, which are, in fact, exclusive cells of human or simian origin.

An important characteristic of herpesviruses is the architecture of the virion. Its size varies between 120 and 300 nm and is described to have a polygonal or round shape with a clearly visible central dot [2, 3]. Until now, it is not exactly known, how many polypeptides are involved in the assembly of the virion, but an average of 30–35 is reported. The virion is structured by four distinct components: envelope, tegument, capsid and core with the genome (fig. 1a).

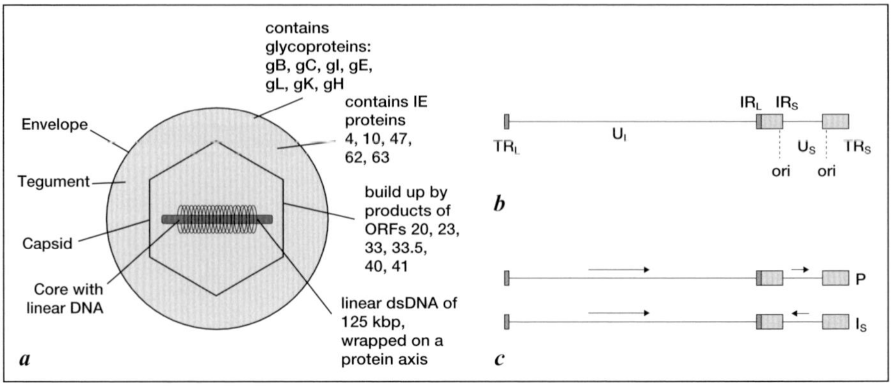

Fig. 1. Schematic drawing of the VZV. ***a*** Overview of its morphology: the important elements – envelope, tegument, capsid and core – are indicated on the left-hand side, important components of each element are given on the right-hand side. ***b*** General structure of the VZV genome. The genome (124,884 bp) can be divided into the unique long (U_L) and unique short (U_S) region, which are flanked by terminal repeats long and short (TR_L, TR_S) and internal repeat long and short (IR_L, IR_S). The origins of DNA replication are located in IR_S and TR_S. ***c*** Isomeric forms of VZV DNA: The P and I_S isomers make up more than 95% of the packaged VZV DNA.

The outer covering envelope has a typical trilaminar appearance [4]. It consists of different membranous elements captured during the transport of the new particles through the nuclear membrane network, Golgi apparatus, rough endoplasmic reticulum, cytoplasmic vesicles and cell surface elements [4–7]. The envelope is interspersed by spikes made up of viral glycoproteins. The VZV genome encodes glycoproteins gB, gC, gE, gH, gI, gK and gL as well as the putative glycoproteins gM and gN [8–10]. The enveloped particles have a final diameter of 180–200 nm and a pleomorphic to spherical shape.

The next inside layer, located directly underneath the envelope, is the tegument [11]. It does not have any distinct properties, but its thickness seems to be variable: virions located in cytoplasmic vacuoles obviously have a thicker tegument than those located in the perinuclear space [12]. The proteins encoded by the open reading frames (ORF) 4, 10, 47, 62 and 63 are found inside the tegument [13, 14]. The tegument surrounds the nucleocapsid.

The nucleocapsid has an icosahedric shape of 100–110 nm in diameter. It is composed of exactly 162 capsomers. Due to the morphology of this capsid structure, it is not possible to distinguish between members of the Herpesviridae. All capsomers occur in a 5:3:2 axial symmetry in which pentameric proteins form the vertices of an 80–120 nm icosahedron. The facets are comprised by hexameric elements [15]. The capsid is built up by the proteins encoded by ORFs 20, 23, 33, 33.5, 40 and 41 (fig. 1a) [9].

The VZV genome is located inside the nucleocapsid. The DNA is coiled upon a protein axis. This combination of linear DNA and proteins is called core [16]. The genome is a linear double stranded DNA molecule of approximately 125 kbp in length and an average G–C content of 46%. This is the smallest genome known in the family of herpesviruses. During transition of the DNA from the capsid into the nucleus of an infected cell, it changes from a linear state into a circular one. It contains at least 69 unique ORFs and three duplicated genes (ORFs 62–71, 63–70 and 64–69, for overview see [10]). The VZV genome consists of two covalently linked segments, U_L and U_S (long, L and short, S), which are composed of unique sequences. Both of these unique segments are flanked by inverted repeat sequences: U_L by IR_L (internal repeat long) and TR_L (terminal repeat long), U_S by IR_S (internal repeat short) and TR_S (terminal repeat short) (fig. 1b). In the genome of the VZV strain Dumas, which is completely sequenced [17], the U_L element has a length of 104,836 bp flanked by 88 bp inverted repeats and the U_S region, which is 5,232 bp in length, is surrounded by inverted repeats of 7,319 bp.

VZV DNA isolated from purified virions can be found in two predominant isomeric forms designated as P (prototype) and I_S (showing an inverted U_S region) (fig. 1c) [18–22]. Other isomeric forms can only be found at very low levels representing 2–5% of the virion DNA. DNA purified from VZV nucleocapsids is infectious as it was first demonstrated by Dumas et al. [23].

The VZV genome contains two origins of replication (ori) [24, 25]. These elements, consisting of a 46 bp palindromic sequence which centres are composed of 16 TA dinucleotide repeats, are located within the inverted repeats flanking the U_S region (fig. 1b). Three internal elements inside these ori-sequences, designated A, B and C, are recognised by the viral origin binding protein encoded by ORF 51 [26].

The Replication Cycle of VZV

The replication cycle of VZV is divided into three different phases: (i) virus adsorption and entry, uncoating, transportation of the capsid to the nucleus and release of the viral DNA into it, (ii) viral gene transcription and translation as well as synthesis of viral DNA and (iii) assembly of new virions, enveloping and egress.

The replication cycle begins when the virus adsorbs to its specific receptors on the surface of the target cell. The adsorption is mediated by viral glycoproteins, the receptors have not yet been precisely identified. However, recent data indicate that the mannose 6-phosphate receptor plays a major role during attachment since at least four VZV envelope glycoproteins contain mannose 6-phosphate [27, and references there in]. After fusion of the viral envelope and the cellular membrane, capsid and tegument proteins are released into the cytoplasm. The capsid is transported to the nuclear pores and releases its nucleic acids by an unknown mechanism. With regard to this process, it is noteworthy that the cytoskeletal architecture of the host cell was found to be altered after infection. Microfilaments and microtubules were subject to reorganisation, while intermediate filaments remained unaffected. These data support the thesis that cellular filament systems play an important role in the transport of virions or nucleocapsids as it is known from HSV [28].

The following expression of viral genes runs according to a very precise cascade. Immediate-early genes (IE; ORFs 4, 61, 62, 63; [29–38]) are transcribed first within a few hours of infection in the absence of de novo protein synthesis. The IE proteins have regulatory functions on the subsequent gene transcription.

Next to the virus-encoded transactivator proteins, cellular transcription factors are also involved in the regulation of VZV gene expression. Most VZV promoters contain *cis*-acting elements which are recognised by ubiquitously expressed cellular factors. The bi-directional promoter of the ORFs 28 and 29 is activated by cooperation of cellular upstream stimulatory factor and the major transactivator protein encoded by ORF 62 (IE62) [39–41]. Other cellular factors of importance are Sp1 and Ap1. Sp1 is one essential factor for the trans-regulation of the activating upstream sequence-element inside the viral glycoprotein I promoter [42] as well as it is implicated in the regulation of the viral glycoprotein E expression by substituting the TATA-box binding protein to initiate transcription [43, 44]. The expression and activation of Ap1 increased significantly after infection of cells with VZV and a knockout of this factor leads to a significant decrease of virus replication [45]. To achieve AP-1 activation, VZV takes advantage of pre-existing cellular signalling pathways such as the MAPK cascades [46]. The ORF61 protein has been demonstrated to be involved in the regulation of this pathway [47].

The induction of transcription of a secondary class of genes, named early-(E) genes, which can be translated into early proteins before the onset of viral DNA replication is dependent on the cooperation of viral IE proteins and cellular transcription factors. Almost all E genes encode proteins with enzymatic properties involved in the replication of viral DNA, like the DNA polymerase (ORF 28), the polymerase processivity factor (ORF 16), the helicase (ORF 55), the primase (ORF 6), the helicase/primase accessory factor (ORF 52), the single-strand DNA binding factor (ORF 29) and origin binding protein (ORF 51).

The VZV DNA replication process itself can be divided into different steps [48]. At first, the linear viral DNA circularises followed by the start of the replication process, which involves the rolling circle-mechanism leading to the formation of head-to-tail concatemers [24]. Isomerisation may occur by homologous recombination between the inverted repeats. Finally, the concatemers are cleaved to generate linear DNA which is packaged into virions.

After DNA replication has begun, late (L) genes are transcribed. Proteins belonging into the group of L products are the glycoproteins as well as those proteins that build up the virus particles.

Due to the aim to achieve a strict and efficient expression of all classes of genes and to repress an up-come of host defence mechanisms, VZV mediates a process known as host shut-off which results in the degradation of cellular mRNA. In contrast to HSV-1, the VZV mediated shut-off is not an immediate-early process but a delayed one. The ORF17 protein, which is the homologue to the HSV virion host shut-off (*vhs*) factor U_L41, is not the main actor to gain the shut-off [49, 50]. Due to its transrepressing properties, recent reports indicate a role of IE63 in putting on the shut-off effect [51, 52].

The degradation of mRNA includes also transcripts of VZV IE genes what is thought to be a part of the switching process from the IE to the E and L gene transcription during the replication cascade [53]. Viral E and L transcripts are also degraded as a consequence of the shut-off [53, 54]. However, evidence is increasing that a broad range of cellular genes are not influenced by the shut-off.

In addition to the host shut-off as a mechanism against host defence, VZV is also capable to prevent the induction of interferon-stimulated anti-viral systems such as PKR and RNase L [55, 56].

After the expression of all three classes of genes has occurred, the newly replicated genomes are wrapped on the protein core, packed inside the newly synthesised capsids and transported outside the host cell. It is still not definitely clear how and in which form the nucleocapsids are transported out of the nucleus and towards the egress. Different hypotheses are still proposed. A widely accepted model is that the capsids get a temporary envelope gained from inner nuclear membranes while entering the perinuclear space. These newly formed particles reach the lumen of the rough endoplasmatic reticulum (rER). The envelope fuses with the rER membrane. The processes resulting in this temporary enveloping at the inner nuclear membrane and the fusion with the rER membrane are not understood yet. Further hypotheses give unknown functions of some glycoproteins in these events. Following this dis-envelopment, naked particles bud into large cytoplasmic vesicles. The viral glycoproteins are released from the trans-Golgi network in additional vesicles that fuse with the cytoplasmic vesicles prior to virion formation. The assembly of fully enveloped virions with functional glycoproteins occurs in these vesicles while they are forwarded to the cell surface. The viral particles are released by exocytosis [8, 57–59]. According to another scenario, cytosolic capsids are wrapped by cisternae of the trans-Golgi network which already contain the glycoproteins. The tegument is thought to bind to those glycoproteins [60].

References

1 Rahaus M, Schunemann S, Wolff MH: Morphological and biological characteristics of varicella-zoster virus in Wolff MH, Schünemann S, Schmidt A (eds): Varicella-zoster Virus: Molecular Biology, Pathogenesis, and Clinical Aspects. Contributions to Microbiology. Basel, Karger, 1999, vol 3, pp 1–9.

2 Ruska H: Ziele und Erfolge der Übermikroskopie in der medizinischen Forschung. Scientia 1943;37:16.

3 Ruska H: Über das Virus der Varizellen und des Zoster. Wien Klin Wochenschr 1943;22:703–704.

4 Epstein MA: Observations on the mode of release of herpes virus from infected HeLa cells. J Cell Biol 1962;12:589–597.

5 Achong BG, Meurisse EV: Observations on the fine structure and replication of varicella virus in cultivated human amnion cells. J Gen Virol 1968;3:305–308.

6 Cok ML, Stevens JG: Replication of varicella-zoste virus in cell culture: an ultrastructural study. J Ultrastruct Res 1970;32:334–350.

7 Stackpole CW: Herpes-type virus of the frog renal adenocarcinoma. I. Virus development in tumor transplants maintained at low temperature. J Virol 1969;4:75–93.

8 Grose C: Glycoproteins encoded by varicella-zoster virus: biosynthesis, phosphorylation, and intracellular trafficking. Annu Rev Microbiol 1990;44:59–80.

9 Kinchington PR, Cohen JI: Viral proteins; in Arvin AM, Gershon AA (eds): Varicella-zoster Virus, Virology and Clinical Managment. Cambridge, Cambridge University Press, 2003, pp 74–104.

10 Ruyechan WT, Hay J: DNA replication; in Arvin AM, Gershon AA (eds): Varicella-zoster Virus, Virology and Clinical Management. Cambridge, Cambridge University Press, 2003, pp 51–73.

11 Roizman B, Furlong D: The replication of herpes viruses; in Fraenkel-Conrat H, Wagner RR (eds): Comprehensive Virology. New York, Plenum, 1974, vol 3, pp 229–403.

12 Falke D, Siegert R, Vogell W: Elektronen-mikroskopische Befunde zur Frage der Doppelmembranbildung des Herpes-simplex Virus. Arch Gesamte Virusforsch 1959;9:484–496.

13 Besser J, Ikoma M, Fabel K, Sommer MH, Zerboni L, Grose C, Arvin AM: Differential requirement for cell fusion and virion formation in the pathogenesis of varicella-zoster virus infection in skin and T cells. J Virol 2004;78:13293–13305.

14 Kinchington PR, Hougland JK, Arvin AM, Ruyechan WT, Hay J: The varicella-zoster virus immediate-early protein IE62 is a major component of virus particles. J Virol 1992;66: 359–366.

15 Almeida JD, Howatson AF, Williams MG: Morphology of varicella (chicken pox) virus. Virology 1962;16:353–355.

16 Puvion-Dutilleul F, Pichard E, Laithier M, Leduc EH: Effect of dehydrating agents on DNA organization in herpes viruses. J Histochem Cytochem 1987;35:635–645.

17 Davison AJ, Scott JE: The complete DNA sequence of varicella-zoster virus. J Gen Virol 1986;67(pt 9):1759–1816.

18 Davison AJ, Scott JE: Molecular cloning of the varicella-zoster virus genome and derivation of six restriction endonuclease maps. J Gen Virol 1983;64(pt 8):1811–1814.

19 Dumas AM, Geelen JL, Weststrate MW, Wertheim P, van der Noordaa J: XbaI, PstI, and BglII restriction enzyme maps of the two orientations of the varicella-zoster virus genome. J Virol 1981;39:390–400.

20 Ecker JR, Hyman RW: Varicella zoster virus DNA exists as two isomers. Proc Natl Acad Sci USA 1982;79:156–160.

21 Straus SE, Aulakh HS, Ruyechan WT, Hay J, Casey TA, Vande Woude GF, Owens J, Smith HA: Structure of varicella-zoster virus DNA. J Virol 1981;40:516–525.

22 Straus SE, Owens J, Ruyechan WT, Takiff HE, Casey TA, Vande Woude GF, Hay J: Molecular cloning and physical mapping of varicella-zoster virus DNA. Proc Natl Acad Sci USA 1982;79: 993–997.
23 Dumas AM, Geelen JL, Maris W, Van der Noordaa J: Infectivity and molecular weight of varicella-zoster virus DNA. J Gen Virol 1980:47.
24 Stow ND, Davison AJ: Identification of a varicella-zoster virus origin of DNA replication and its activation by herpes simplex virus type 1 gene products. J Gen Virol 1986;67(pt 8): 1613–1623.
25 Stow ND, Weir HM, Stow EC: Analysis of the binding sites for the varicella-zoster virus gene 51 product within the viral origin of DNA replication. Virology 1990;177:570–577.
26 Koff A, Tegtmeyer P: Characterization of major recognition sequences for a herpes simplex virus type 1 origin-binding protein. J Virol 1988;62:4096–4103.
27 Chen JJ, Zhu Z, Gershon AA, Gershon MD: Mannose 6-phosphate receptor dependence of varicella zoster virus infection in vitro and in the epidermis during varicella and zoster. Cell 2004;119: 915–926.
28 Kuhn M, Desloges N, Rahaus M, Wolff MH: Varicella-zoster virus infection influences expression and organization of actin and alpha-tubulin but does not affect lamin A and vimentin. Intervirology 2005;48:312–320.
29 Debrus S, Sadzot-Delvaux C, Nikkels AF, Piette J, Rentier B: Varicella-zoster virus gene 63 encodes an immediate-early protein that is abundantly expressed during latency. J Virol 1995;69:3240–3245.
30 Defechereux P, Debrus S, Baudoux L, Rentier B, Piette J: Varicella-zoster virus open reading frame 4 encodes an immediate-early protein with posttranscriptional regulatory properties. J Virol 1997;71:7073–7079.
31 Disney GH, McKee TA, Preston CM, Everett RD: The product of varicella-zoster virus gene 62 autoregulates its own promoter. J Gen Virol 1990;71:2999–3003.
32 Inchauspe G, Nagpal S, Ostrove JM: Mapping of two varicella-zoster virus-encoded genes that activate the expression of viral early and late genes. Virology 1989;173:700–709.
33 Kinchington PR, Bookey D, Turse SE: The transcriptional regulatory proteins encoded by varicella-zoster virus open reading frames (ORFs) 4 and 63, but not ORF 61, are associated with purified virus particles. J Virol 1995;69:4274–4282.
34 Kinchington PR, Vergnes JP, Turse SE: Transcriptional mapping of varicella-zoster virus regulatory proteins. Neurology 1995;45:S33–35.
35 Moriuchi M, Moriuchi H, Straus SE, Cohen JI: Varicella-zoster virus (VZV) virion-associated transactivator open reading frame 62 protein enhances the infectivity of VZV DNA. Virology 1994;200:297–300.
36 Ng TI, Keenan L, Kinchington PR, Grose C: Phosphorylation of varicella-zoster virus open reading frame (ORF) 62 regulatory product by viral ORF 47-associated protein kinase. J Virol 1994;68:1350–1359.
37 Stevenson D, Colman KL, Davison AJ: Characterization of the varicella-zoster virus gene 61 protein. J Gen Virol 1992;73:521–530.
38 Stevenson D, Xue M, Hay J, Ruyechan WT: Phosphorylation and nuclear localization of the varicella-zoster virus gene 63 protein. J Virol 1996;70:658–662.
39 Meier JL, Luo X, Sawadogo M, Straus SE: The cellular transcription factor USF cooperates with varicella-zoster virus immediate-early protein 62 to symmetrically activate a bidirectional viral promoter. Mol Cell Biol 1994;14:6896–6906.
40 Rahaus M, Desloges N, Yang M, Ruyechan WT, Wolff MH: Transcription factor USF, expressed during the entire phase of varicella-zoster virus infection, interacts physically with the major viral transactivator IE62 and plays a significant role in virus replication. J Gen Virol 2003;84: 2957–2967.
41 Yang M, Hay J, Ruyechan WT: The DNA element controlling expression of the varicella-zoster virus open reading frame 28 and 29 genes consists of two divergent unidirectional promoters which have a common USF site. J Virol 2004;78:10939–10952.
42 Ito H, Sommer MH, Zerboni L, He H, Boucaud D, Hay J, Ruyechan W, Arvin AM: Promoter sequences of varicella-zoster virus glycoprotein I targeted by cellular transactivating factors Sp1 and USF determine virulence in skin and T cells in SCIDhu mice in vivo. J Virol 2003;77: 489–498.

43 Rahaus M, Wolff MH: Influence of different cellular transcription factors on the regulation of varicella-zoster virus glycoproteins E (gE) and I (gI) UTR's activity. Virus Res 1999;62:77–88.
44 Rahaus M, Wolff MH: Transcription factor Sp1 is involved in the regulation of varicella-zoster virus glycoprotein E. Virus Res 2000;69:69–81.
45 Rahaus M, Wolff MH: Reciprocal effects of varicella-zoster virus (VZV) and AP1: activation of jun, fos and ATF-2 after VZV infection and their importance for the regulation of viral genes. Virus Res 2003;92:9–21.
46 Rahaus M, Desloges N, Wolff MH: Replication of varicella-zoster virus is influenced by the levels of JNK/SAPK and p38/MAPK activation. J Gen Virol 2004;85:3529–3540.
47 Rahaus M, Desloges N, Wolff MH: ORF61 protein of Varicella-zoster virus influences JNK/SAPK and p38/MAPK phosphorylation. J Med Virol 2005;76:424–433.
48 Cohen JI, Straus SE: Varicella-zoster virus and its replication; in Fields BN, Knipe DM, Howley PM (eds): Field's Virology. Philadelphia, Lippincott-Raven, 1996, vol 2, pp 2525–2545.
49 Sato H, Callanan LD, Pesnicak L, Krogmann T, Cohen JI: Varicella-zoster virus (VZV) ORF17 protein induces RNA cleavage and is critical for replication of VZV at 37 degrees C but not 33 degrees C. J Virol 2002;76:11012–11023.
50 Waterboer T, Rahaus M, Wolff MH: Varicella-zoster virus (VZV) mediates a delayed host shutoff independent of open reading frame (ORF) 17 expression. Virus Genes 2002;24:49–56.
51 Desloges N, Rahaus M, Wolff MH: The varicella-zoster virus-mediated delayed host shutoff: open reading frame 17 has no major function, whereas immediate-early 63 protein represses heterologous gene expression. Microbes Infect 2005;7:1519–1529.
52 Di Valentin E, Bontems S, Habran L, Jolois O, Markine-Goriaynoff N, Vanderplasschen A, Sadzot-Delvaux C, Piette J: Varicella-zoster virus IE63 protein represses the basal transcription machinery by disorganizing the pre-initiation complex. Biol Chem 2005;386:255–267.
53 Hamouda T, McPhee R, Hsia SC, Read GS, Holland TC, King SR: Inhibition of human immunodeficiency virus replication by the herpes simplex virus virion host shut-off protein. J Virol 1997;71:5521–5527.
54 Oroskar AA, Read GS: Control of mRNA stability by the virion host shutoff function of herpes simplex virus. J Virol 1989;63:1897–1906.
55 Desloges N, Rahaus M, Wolff MH: Varicella-zoster virus does not significantly induce the cell defence mechanism mediated by the 2–5A/RNase L pathway during its replication cycle. Med Microbiol Immunol (Berl) 2004;194:25–31.
56 Desloges N, Rahaus M, Wolff MH: Role of the protein kinase PKR in the inhibition of varicella-zoster virus replication by beta interferon and gamma interferon. J Gen Virol 2005;86: 1–6.
57 Gershon MD, Gershon AA: Role of glycoproteins in varicella-zoster virus infection; in Wolff MH, Schünemann S, Schmidt A (eds): Varicella-zoster Virus, Molecular Biology, Pathogenesis, and Clinical Aspects. Contributions to Microbiology. Basel, Karger, 1999, vol 3, pp 43–60.
58 Jones F, Grose C: Role of cytoplasmic vacuoles in varicella-zoster virus glycoprotein trafficking and virion envelopment. J Virol 1988;62:2701–2711.
59 Sadzot-Delvaux C, Baudoux L, Defechereux P, Piette J, Rentier B: Overview of the replication cycle of varicella-zoster virus; in Wolff MH, Schünemann S, Schmidt A (eds): Varicella-zoster Virus: Molecular Biology, Pathogenesis and Clinical Aspects. Contributions to Microbiology. Basel, Karger, 1999, vol 3, pp 21–42.
60 Gershon AA, Sherman DL, Zhu Z, Gabel CA, Ambron RT, Gershon MD: Intracellular transport of newly synthesized varicella-zoster virus: final envelopment in the trans-Golgi network. J Virol 1994;68:6372–6390.

Manfred H. Wolff
Institute of Microbiology and Virology
Private University of Witten-Herdecke
Stockumer Strasse 10
DE–58448 Witten (Germany)
Tel. +49 2302 926107, Fax +49 2302 926220, E-Mail mhwolff@uni-wh.de

Gross G, Doerr HW (eds): Herpes Zoster.
Monogr Virol. Basel, Karger, 2006, vol 26, pp 9–12

Latency and Reactivation of VZV

Octavian Lungu[a], *Anne A. Gershon*[b]

Departments of [a]Microbiology and [b]Pediatrics, Columbia University, College of Physicians and Surgeons, New York, N.Y., USA

Primary, Latent and Reactivated VZV Infection

Like other herpes viruses, VZV is produced following a temporally coordinated expression of its genes. Upon entering a naïve cell viral DNA and virion-associated proteins are transported to the cell nucleus and immediate early (IE) genes are transcribed. After the corresponding mRNAs are translated in the cell cytoplasm, the specific IE proteins return to the nucleus to initiate the expression of early (E) genes which are involved in viral replication. Then, IE and E gene products prompt the expression of late (L) genes whose products are required for packaging of the newly synthesized viral DNA and assembly of mature virions. Virus then exits the cell to infect adjacent uninfected cells. Complete or partial interruption of this regulated cycle of virus gene expression results in latency. The re-establishment of the co-ordinated sequence of gene expression results in reactivation.

Wild-type VZV infects varicella-susceptible individuals through the respiratory route. After an initial replication in the respiratory tract lymphoid tissue, a primary viremia occurs, then the virus reaches the central lymphoid system where it continues to replicate. Due to a subsequent secondary viremia of larger magnitude than the first, the skin is productively infected and the vesicular rash and other clinical signs of chickenpox are apparent. During chickenpox VZV DNA and proteins expressed by IE, E and L genes are detected in both the nucleus and the cytoplasm of the infected cells (fig. 1, lytic VZV Varicella Skin column).

During primary infection through retrograde neuronal transport from the skin and possibly hematogeneuos spread, or both routes, the virus reaches dorsal root ganglia (DRG) and establishes latent infection in the sensory neurons and the supporting non-neuronal ganglionic cells [3, 4] (fig. 1, latent VZV DNA DRG panel). Note that viral DNA is present in the nucleus of a neuron

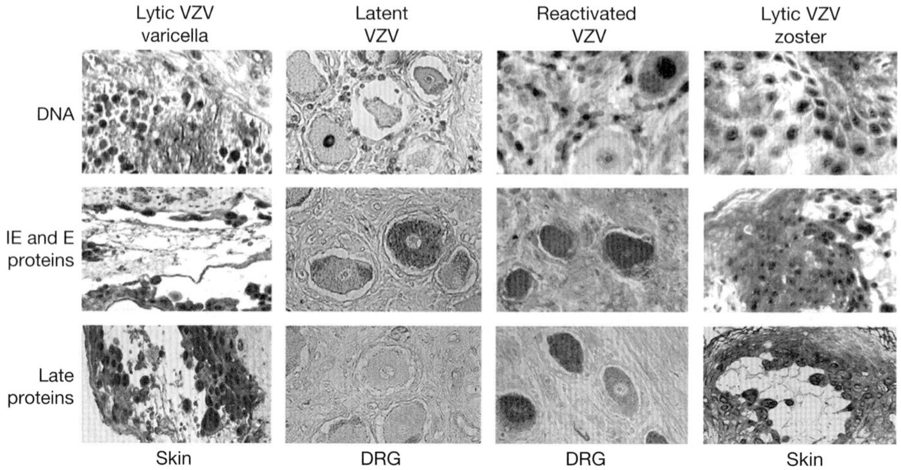

Fig. 1. VZV DNA and proteins identified by in situ hybridization and immunohistochemistry during VZV primary infection or VZV latency and reactivation. The presence of virus DNA or proteins (IE63p, L68/gE – as examples) is indicated by a dark colouration of cells nuclei, cytoplasm or both.

and several satellite cells. As opposed to the closely related herpes simplex viruses, which are silent during their latency in DRG, VZV expresses at least 5 IE and E genes while in the DRG. Transcripts encoded by IE 4, 62, 63 and E 21, 29 VZV genes [5, 6] and the corresponding regulatory proteins translated by these transcripts [2, 7] can be identified in latently infected DRG. Late VZV genes are not expressed during latency and L proteins are not detected in latently infected DRG [2] (fig. 1, Latent VZV Late proteins DRG panel). IE and E proteins localize to the cytoplasm of neurons during latency [2] (fig. 1, latent VZV IE and E proteins DRG panel). The exclusion of IE and E proteins from the cell nucleus, where they perform their regulatory functions during productive infection, is a hallmark of VZV latency and is thought to be a crucial factor for the maintenance of virus latency.

For reasons yet to be defined, VZV can exit its latent state to cause a productive infection in the infected ganglia. During reactivation VZV IE and E proteins reach the nuclei of cells [2] (fig. 1, reactivated VZV IE and E proteins DRG panel), re-establishing the co-ordinated cycle of VZV genes expression that results in the production of L proteins [2] (fig. 1, reactivated VZV Late proteins DRG panel) and lytic infection. The mature virions identified in the cytoplasm of neurons as VZV DNA [2] (fig. 1, reactivated VZV DNA DRG panel – light dark staining in the cytoplasm of a large cell) then use anterograde neuronal transport to reach the skin where they manifest as shingles. As in chickenpox, during shingles, VZV DNA and proteins expressed from all kinetic classes genes are detected in both the nucleus and the cytoplasm of the infected cells (fig. 1, lytic VZV Zoster Skin column).

Conclusions

The puzzle of VZV latency and reactivation remains incompletely understood. It appears that immunity to VZV, and the aberrant intracellular localization of VZV regulatory proteins during virus latency in DRG are crucial in determining if VZV remains latent or reactivates in the sensory ganglia. The recent development of an in vitro model of VZV latency in guinea pig enteric neurons [8], that closely mimics VZV latency in human DRG, will allow identification of the viral and cellular factors that influence the state of VZV infection in humans, and will contribute to the understanding of the molecular mechanisms that govern VZV latency and reactivation. In addition, the latency of the VZV vaccine (Oka) can also be studied using this in vitro model of VZV infection, allowing the determination of the reactivation potential of the VZV vaccine when compared with wild-type VZV.

Acknowledgements

Studies done by the authors were supported by grants AI124021 and AI27187 from the US Public Health Service to Anne Gershon and Saul Silverstein.

References

1 Straus SE: Shingles-sorrows, salves, and solutions. JAMA 1993;269:1836–1839.
2 Lungu O, Annunziato PW, Gershon A, Staugaitis SM, Josefson D, LaRussa P, Silverstein SJ: Reactivated and latent varicella-zoster virus in human dorsal root ganglia. Proc Natl Acad Sci USA 1995;92:10980–10984.
3 Cohrs RJ, Barbour M, Gilden DH: Varicella-zoster virus (VZV) transcription during latency in human ganglia: detection of transcripts mapping to genes 21, 29, 62, and 63 in a cDNA library enriched for VZV RNA. J Virol 1996;70:2789–2796.
4 Meier JL, Holman RP, Croen KD, Smialek JE, Straus SE: Varicella-zoster virus transcription in human trigeminal ganglia. Virology 1993;193:193–200.
5 Kennedy PGE, Grinfeld E, Gow JW: Latent varicella-zoster virus is located predominantly in neurons in human trigeminal ganglia. Proc Natl Acad Sci USA 1998;95:4658–4662.
6 Lungu O, Annunziato PW, Gershon A, Staugaitis SM, Josefson D, LaRussa P, Silverstein SJ: Reactivated and latent varicella-zoster virus in human dorsal root ganglia. Proc Natl Acad Sci USA 1995;92:10980–10984.

7 Mahalingam R, Wellish M, Cohrs R, Debrus S, Piette J, Rentier B, Gilden DH: Expression of protein encoded by varicella-zoster virus open reading frame 63 in latently infected human ganglionic neurons. Proc Natl Acad Sci USA 1996;93:2122–2124.
8 Chen JJ, Gershon AA, Li ZS, Lungu O, Silverstein S, Gershon MD: Latent and lytic infection of isolated guinea pig enteric ganglia by varicella zoster virus. J Med Virol 2003;70(suppl 1):S71–S78.

Octavian Lungu
Department of Pediatrics, Columbia University
College of Physicians and Surgeons
701 West 168th Street
New York, NY 10032 (USA)
Tel. +1 212 305 4137, Fax +1 212 305 5106, E-Mail ol4@columbia.edu

Gross G, Doerr HW (eds): Herpes Zoster.
Monogr Virol. Basel, Karger, 2006, vol 26, pp 13–19

Laboratory Confirmation of Herpes Zoster

B. Ludwig, S. Buxbaum, H.W. Doerr

Institute of Virology, University of Frankfurt, Frankfurt am Main, Germany

The diagnosis of herpes zoster is usually a clinical diagnosis and therefore laboratory confirmation is not necessary. In cases where herpes zoster is in question a rapid laboratory diagnosis may be helpful, especially in immunocompromised patients, when antiviral therapy must be initiated as soon as possible. Similarly, laboratory diagnosis may be useful for the determination of the antibody status and the diagnostic clarification of ophthalmic and central nerve system complications of a VZV infection. Serology is a helpful tool to recognise VZV seronegative individuals having a risk to acquire an infection (e.g. non-immune women during the first third of the pregnancy). Generally, we can distinguish between the direct prove of the pathogen itself and the immunological reaction of the infected person (humoral or cellular immunity). Today virological methods for the direct detection of infectious virus, viral DNA or viral proteins are electron microscopy, cell culture, immunofluorescence assays and polymerase chain reaction (PCR) (figs. 1–3). Serological diagnosis of VZV infection/reactivation covers a number of different methods. Detection of cellular immunity is difficult to carry out and has therefore no role in routine diagnosis so far.

Direct Examination of Clinical Material

Microscopy

Electron microscopy (EM) can be used, if a rapid diagnosis is needed or differentiation of VZV and smallpox is required. Typical herpesvirus particles can be detected in the fluid of early vesicles or in infected cells scraped from the ground of a lesion. Since the viruses of the herpes group have all the same morphology, EM does not differentiate between VZV and HSV unless

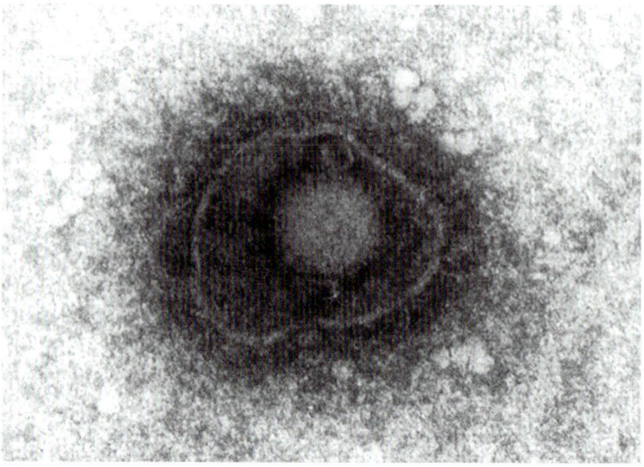

Fig. 1. VZV vesicles: electron microscopy.

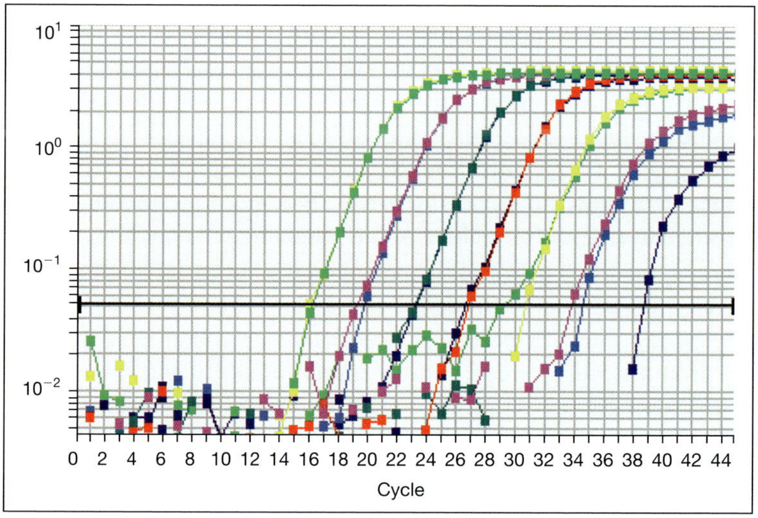

Fig. 2. PCR investigation by real-time technique.

combined with further immunological techniques [1–3] which are normally not used in routine diagnosis (fig. 1).

Direct Immunofluorescence Assay

Direct immunofluorescent antibody staining with monoclonal antibodies allows a rapid and specific diagnosis of VZV suspected lesions. Even cells from

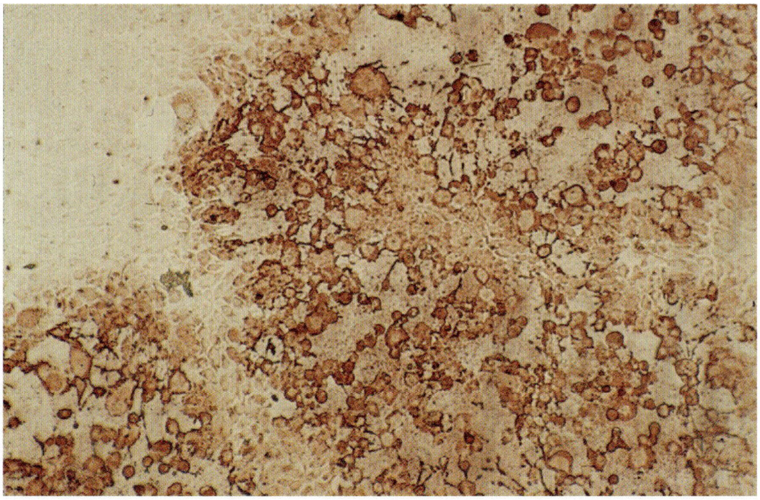

Fig. 3. VZV-specific cytopathologic effect in retinal pigment epithelial cells.

crusted lesions contain enough viral antigen for detection. Some investigators recommend this assay as 'method of choice' for the diagnosis of VZV-infection, because it is easy to achieve and more sensitive than virus isolation and serology [4–8]. To reach optimal sensitivity cells have to be scraped off from the base of a lesion after unroofing the fresh vesicle. The sensitivity and recovery of this method depends on the quality of the scrap of tissue. Before immunohistostaining was available, cytologic methods (Tyzzer or Tzank smears) for detection of multinucleate giant cells with intranuclear inclusions were used. Like EM these methods are depending on the investigator's experience and allow no differentiation between HSV and VZV [9].

Detection of Viral DNA

The most sensitive and specific methods for detection of VZV in clinical specimens (like vitreous fluid or cerebrospinal fluid) are molecular amplification methods, especially the PCR [10–13]. PCR has proven to be a valuable tool for the diagnosis of VZV disease of the central nerve system, the ophthalmic division and VZV associated vasculopathy [14–16]. Further, it is also useful for the detection of VZV genome in vesicle fluid and crusts, especially when antiviral therapy has been started before [17]. Detection of VZV genome does not prove that infectious virus is present, and therefore detection of VZV with this highly sensitive method has to be correlated with the clinical condition before an aetiologic role is assumed [18]. Using quantitative PCR (fig. 2) it is possibly to find low numbers of genome copies in latently infected ganglionic cells without showing infectious virus forms or subsequent disease [14, 19, 20]. For special cases a molecular characterisation of VZV isolates can be ruled out by DNA sequencing, which allows to ascertain relatedness of different clinical isolates and detection of some acyclovir resistant mutations [21, 22].

Virus Isolation

Virus isolation in cell culture remains, in cases of positive testing, the most reliable method for the proof of infectious virus. It is less sensitive than direct immunofluorescent antibody staining, since viable virus persists shorter in vesicles and is more labile than viral antigen [23]. Usually results of viral cultures cannot be obtained fast enough to influence clinical decisions, but they can confirm diagnosis of VZV infection made by other methods, e.g. the antigen assay [24]. In addition, drug resistance of virus can be ruled out by testing cultured virus [25]. Thereby, it should be noted that isolation of VZV strains, especially acyclovir resistant strains and their characterisation are difficult by tissue culture procedures. This is mainly because VZV is not stable, highly cell associated and replicating slowly in cell culture of low titres [23, 26].

Infectious VZV is usually recoverable from vesicle fluid or the base of a fresh zoster lesions for up to 7 days [24, 27]. A spectrum of cultured cells, either of primary or of established cell lines can be used for the isolation and growth of VZV (fig. 3). The most recommended cell cultures are either primary or low passage cultures of human fibroblasts derived from embryonic skin, lung tissue, or preputial tissues. Even under improved conditions, a significant percentage of viral cultures remains negative, despite positive VZV-antigen or -DNA detection [4, 10, 17]. Methods like 'shell vial' cultures with centrifugation and VZV protein staining improve the sensitivity and allow a more rapid identification of positive specimens [28, 29]. Positive results may be available within 1–3 days after inoculation of cell culture [30, 31], while in conventional cultures up to 21 days are required [27].

Determination of the Immunity Status

Serological Diagnosis

Serological assays are frequently used for the diagnosis of chickenpox and herpes zoster. The preferred methods are enzyme immunoassays (ELISA) which show both a high specificity and sensitivity [32]. However, the detection of VZV-specific antibodies for the diagnosis of atypical zoster is only of limited value when a rapid confirmation of VZV-infection is needed. Nevertheless, some problems remain in the serological diagnosis of VZV infections. VZV reactivation induces often a significant rise of IgG and IgA [33, 34] antibodies, which are found in 50–60% [10, 35] of the patients. However IgM antibodies are also sometimes detectable. Thus differentiation of primary and recurrent infection can be difficult [36]. Furthermore, sharing of antigens between VZV and HSV can result in high anti HSV and anti VZV levels and it will be difficult to differentiate between both diseases without further information or additional tests [23, 37]. The main value of serologic assays is the determination of the immune status of individuals, whose history of chickenpox is unknown [24]. In patients with pain syndromes or facial paresis due to zoster *sine herpete* a rise of VZV-IgG values in consecutive serum samples may be helpful to identify the aetiology [38]. Other serological test systems like the neutralisation assay and the fluorescent-antibody membrane antigen assay are too complex to carry out in routine diagnostic testing [23].

Determination of Cell-Mediated Immunity

The main role of cell-mediated immunity (CMI) is to prevent reactivation and to limit an established infection. Detection of CMI can be carried out by measuring the proliferative response [39] or by detection of cytokine production [40–42]. Examples for such methods are the determination of intracellular cytokines by flow cytometry [43] or of secreted extracellular cytokines by ELISA [44] or ELISPOT [45]. Generally all procedures for measurement of CMI are technically complex, time consuming, labour intensive and therefore used only in special cases on small scales. Nevertheless, in the future the demand for clinical studies concerning VZV-specific CMI on a large scale will rise considerably and so the development of well reproducible and easily practicable diagnosis methods will be necessary.

References

1 Folkers E, Vreeswijk J, Oranje AP, Duivenvoorden JN: Rapid diagnosis in varicella and herpes zoster: re-evaluation of direct smear (Tzanck test) and electron microscopy including colloidal gold immuno-electron microscopy in comparison with virus isolation. Br J Dermatol 1989;121:287–296.

2 Folkers E, Vreeswijk J, Oranje AP, Wagenaar F, Duivenvoorden JN: Improved detection of HSV by electron microscopy in clinical specimens using ultracentrifugation and colloidal gold immunoelectron microscopy: comparison with viral culture and cytodiagnosis. J Virol Methods 1991;34: 273–289.

3 Vreeswijk J, Folkers E, Wagenaar F, Kapsenberg JG: The use of colloidal gold immunoelectron microscopy to diagnose varicella-zoster virus (VZV) infections by rapid discrimination between VZV, HSV-1 and HSV-2. J Virol Methods 1988;22:255–271.

4 Dahl H, Marcoccia J, Linde A: Antigen detection: the method of choice in comparison with virus isolation and serology for laboratory diagnosis of herpes zoster in human immunodeficiency virus-infected patients. J Clin Microbiol 1997;35:347–349.

5 Drew WL, Mintz L: Rapid diagnosis of varicella-zoster virus infection by direct immunofluorescence. Am J Clin Pathol 1980;73:699–701.

6 Gleaves CA, Lee CF, Bustamante CI, Meyers JD: Use of murine monoclonal antibodies for laboratory diagnosis of varicella-zoster virus infection. J Clin Microbiol 1988;26:1623–1625.
7 Perez JL, Garcia A, Niubo J, Salva J, Podzamczer D, Martin R: Comparison of techniques and evaluation of three commercial monoclonal antibodies for laboratory diagnosis of varicella-zoster virus in mucocutaneous specimens. J Clin Microbiol 1994;32:1610–1613.
8 Coffin SE, Hodinka RL: Utility of direct immunofluorescence and virus culture for detection of varicella-zoster virus in skin lesions. J Clin Microbiol 1995;33:2792–2795.
9 Waner JL, Weller TH, Schmidt NJ, Emmons RW (eds): Diagnostic Procedures for Viral, Rickettsial And Chlamydial Infections, ed 6. Varicella-Zoster Virus. Washington, DC, American Public Health Association, 1998, vol 13, pp 379–406.
10 Sauerbrei A, Eichhorn U, Schacke M, Wutzler P: Laboratory diagnosis of herpes zoster. J Clin Virol 1999;14:31–36.
11 Weidmann M, Meyer-Konig U, Hufert FT: Rapid detection of herpes simplex virus and varicella-zoster virus infections by real-time PCR. J Clin Microbiol 2003;41:1565–1568.
12 van Doornum GJ, Guldemeester J, Osterhaus AD, Niesters HG: Diagnosing herpesvirus infections by real-time amplification and rapid culture. J Clin Microbiol 2003;41:576–580.
13 Nahass GT, Goldstein BA, Zhu WY, Serfling U, Penneys NS, Leonardi CL: Comparison of Tzanck smear, viral culture, and DNA diagnostic methods in detection of herpes simplex and varicella-zoster infection. JAMA 1992;268:2541–2544.
14 Kleinschmidt-DeMasters BK, Gilden DH: Varicella-Zoster virus infections of the nervous system: clinical and pathologic correlates. Arch Pathol Lab Med 2001;125:770–780.
15 Madhavan HN, Priya K: The diagnostic significance of enzyme linked immuno-sorbent assay for herpes simplex, varicella zoster and cytomegalovirus retinitis. Indian J Ophthalmol 2003;51: 71–75.
16 Gargiulo F, De Francesco MA, Nascimbeni G, et al: Polymerase chain reaction as a rapid diagnostic tool for therapy of acute retinal necrosis syndrome. J Med Virol 2003;69:397–400.
17 Beards G, Graham C, Pillay D: Investigation of vesicular rashes for HSV and VZV by PCR. J Med Virol 1998;54:155–157.
18 Schunemann S, Mainka C, Wolff MH: Subclinical reactivation of varicella-zoster virus in immunocompromised and immunocompetent individuals. Intervirology 1998;41:98–102.
19 Liedtke W, Opalka B, Zimmermann CW, Lignitz E: Age distribution of latent herpes simplex virus 1 and varicella-zoster virus genome in human nervous tissue. J Neurol Sci 1993;116:6–11.
20 Mahalingam R, Wellish M, Lederer D, Forghani B, Cohrs R, Gilden D: Quantitation of latent varicella-zoster virus DNA in human trigeminal ganglia by polymerase chain reaction. J Virol 1993;67:2381–2384.
21 Morfin F, Thouvenot D, Turenne-Tessier M, Lina B, Aymard M, Ooka T: Phenotypic and genetic characterization of thymidine kinase from clinical strains of varicella-zoster virus resistant to acyclovir. Antimicrob Agents Chemother 1999;43:2412–2416.
22 Boivin G, Edelman CK, Pedneault L, Talarico CL, Biron KK, Balfour HH Jr: Phenotypic and genotypic characterization of acyclovir-resistant varicella-zoster viruses isolated from persons with AIDS. J Infect Dis 1994;170:68–75.
23 Gershon AA, Forghani B, Lennette EH, Lennette DA, Lennette ET (eds): Diagnostic Procedures for Viral, Rickettsial and Chlamydial Infections, ed 7. Varicella-Zoster Virus. Washington, DC, American Public Health Association, 1995, Vol 42, pp 601–613.
24 Arvin AM, Knipe DM, Howley PM (eds): Fields Virology, ed 4. Varicella-Zoster Virus. Philadelphia, Lippincott Willians and Wilkins, 2001, Vol 79, pp 2731–2767.
25 Pillay D, Emery VC, Mutimer D, et al: Guidelines for laboratory monitoring of treatment of persistent virus infections. J Clin Virol 2002;25:73–92.
26 Sahli R, Andrei G, Estrade C, Snoeck R, Meylan PR: A rapid phenotypic assay for detection of acyclovir-resistant varicella-zoster virus with mutations in the thymidine kinase open reading frame. Antimicrob Agents Chemother 2000;44:873–878.
27 Breuer J, Harper DR, Kangro HO, Zuckerman A, Banatvala JE, Pattison JR (eds): Principles and Practice Of Clinical Virology, ed 4. Varicella Zoster. Baffins Lane, Chichester, West Sussex PO19 1UD, England, John Wiley and Sons, 2000, vol 2B, pp 47–78.

28 Schirm J, Meulenberg JJ, Pastoor GW, Voorst Vader PC, Schroder FP: Rapid detection of varicella-zoster virus in clinical specimens using monoclonal antibodies on shell vials and smears. J Med Virol 1989;28:1–6.
29 West PG, Aldrich B, Hartwig R, Haller GJ: Increased detection rate for varicella-zoster virus with combination of two techniques. J Clin Microbiol 1988;26:2680–2681.
30 Brinker JP, Doern GV: Enhancement of varicella-zoster virus detection in A-549 shell vials by use of freeze-thawed specimens, extended incubation, and 'a centrifuged, not incubated' direct detection method. Diagn Microbiol Infect Dis 1998;31:555–558.
31 Brinker JP, Doern GV: Comparison of MRC-5 and A-549 cells in conventional culture tubes and shell vial assays for the detection of varicella-zoster virus. Diagn Microbiol Infect Dis 1993;17:75–77.
32 Weinberg A, Hayward AR, Masters HB, Ogu IA, Levin MJ: Comparison of two methods for detecting varicella-zoster virus antibody with varicella-zoster virus cell-mediated immunity. J Clin Microbiol 1996;34:445–446.
33 Wültzer P, Porstmann T (eds): Virusdiagnostik. Varicella-Zoster-Virus. Berlin, Wien, Blackwell Wissenschafts-Verlag, 1996, pp 291–302.
34 Doerr HW, Rentschler M, Scheifler G: Serologic detection of active infections with human herpes viruses (CMV, EBV, HSV, VZV): diagnostic potential of IgA class and IgG subclass-specific antibodies. Infection 1987;15:93–98.
35 Wittek AE, Arvin AM, Koropchak CM: Serum immunoglobulin A antibody to varicella-zoster virus in subjects with primary varicella and herpes zoster infections and in immune subjects. J Clin Microbiol 1983;18:1146–1149.
36 Arvin AM, Koropchak CM: Immunoglobulins M and G to varicella-zoster virus measured by solid-phase radioimmunoassay: antibody responses to varicella and herpes zoster infections. J Clin Microbiol 1980;12:367–374.
37 Schmidt NJ: Further evidence for common antigens in herpes simplex and varicella-zoster viruses. J Med Virol 1982;9:27–36.
38 Gilden DH, Dueland AN, Devlin ME, Mahalingam R, Cohrs R: Varicella-zoster virus reactivation without rash. J Infect Dis 1992;166(suppl 1):S30–S34.
39 Hayward AR, Zerbe GO, Levin MJ: Clinical application of responder cell frequency estimates with four years of follow up. J Immunol Methods 1994;170:27–36.
40 Zhang Y, Cosyns M, Levin MJ, Hayward AR: Cytokine production in varicella zoster virus-stimulated limiting dilution lymphocyte cultures. Clin Exp Immunol 1994;98:128–133.
41 Zhang Y, White CJ, Levin M, Hayward A: Cytokine production in varicella-zoster virus-stimulated lymphocyte cultures. Neurology 1995;45(suppl 8):S38–S40.
42 Warnatz K, Draeger R, Schlesier M, Peter HH: Successful IL-2 therapy for relapsing herpes zoster infection in a patient with idiopathic CD4+ T lymphocytopenia. Immunobiology 2000;202:204–211.
43 Asanuma H, Sharp M, Maecker HT, Maino VC, Arvin AM: Frequencies of memory T cells specific for varicella-zoster virus, herpes simplex virus, and cytomegalovirus by intracellular detection of cytokine expression. J Infect Dis 2000;181:859–866.
44 Svahn A, Linde A, Thorstensson R, Karlen K, Andersson L, Gaines H: Development and evaluation of a flow-cytometric assay of specific cell-mediated immune response in activated whole blood for the detection of cell-mediated immunity against varicella-zoster virus. J Immunol Methods 2003;277:17–25.
45 Smith JG, Liu X, Kaufhold RM, Clair J, Caulfield MJ: Development and validation of a gamma interferon ELISPOT assay for quantitation of cellular immune responses to varicella-zoster virus. Clin Diagn Lab Immunol 2001;8:871–879.

Professor H.W. Doerr, MD
Institute of Virology, University of Frankfurt
Paul-Ehrlich-Strasse 40
DE–60596 Frankfurt am Main (Germany)
Tel. +49 69 63015219, Fax +49 69 63016477, E-Mail H.W.Doerr@em.uni-frankfurt.de

Gross G, Doerr HW (eds): Herpes Zoster.
Monogr Virol. Basel, Karger, 2006, vol 26, pp 20–36

Clinical Picture and Complications of Herpes Zoster: The View of the Dermatologist

G. Gross

Department of Dermatology and Venereology, University of Rostock, Rostock, Germany

Herpes zoster (shingles, zoster) is a common neurocutaneous disease resulting from reactivation of latent varicella–zoster virus (VZV) infection acquired during primary VZV-infection (varicella, chickenpox). Herpes zoster presents as a painful characteristically unilateral cutaneous rash in the sensory innervation region of a spinal nerve or a cranial nerve. Unlike varicella herpes zoster is a sporadic disease with an estimated lifetime incidence of 10–20% [1]. Whereas, varicella is generally a disease of childhood, herpes zoster becomes more common with increasing age. Factors that decrease immune function, such as human immunodeficiency virus infection, chemotherapy, malignancies and chronic corticosteroid use may also increase the risk of developing herpes zoster. The main risk factor for the development of herpes zoster, however is increasing age, leading to a decline of VZV-specific cell-mediated immunity. Incidence of zoster rises steadily until adulthood and remains constant with about 2–3 cases per 1,000 per year until the end of the fourth decade of life. In persons older than 50 years of age the incidence strongly increases to approximately 5 cases per 1,000 persons per year. Individuals in the sixth to seventh decade have an incidence rate of 6–7 cases per 1,000 and individuals beyond the age of 80 have an incidence of more than 10 cases per 1,000 per year. According to Hope-Simpson [2] more than half of all people who reach 85 years of age will develop herpes zoster at any point of their life.

Persons older than 50 years of age affected by herpes zoster may suffer a significant decrease of quality of life. These persons and immunocompromised individuals of any age are at increased risk for severe complications involving the skin, the eye, internal organs and the peripheral and central nervous system.

About 20% of patients with shingles develop prolonged pain and postherpetic neuralgia (PHN). The most established risk factor for PHN is again age. This complication occurs nearly 50 times more often in patients older than 50 years of age. Other possible risk factors for the development of PHN are ophthalmic zoster, zoster oticus and a history of prodromal pain before appearance of the rash. Growing life expectancy and the increasing number of elderly in Europe has resulted in a higher population risk for herpes zoster and chronic zoster pain. HIV-infected individuals and adults suffering from cancer have a much higher herpes zoster incidence than immunocompetent persons of the same age [3, 4]. The occurrence of herpes zoster in HIV-infected patients, however does not appear to increase the risk of acquired immunodeficiency syndrome (AIDS) and is less dependent on the CD4-count than AIDS-related opportunistic infections [5]. Furthermore, there is no evidence that herpes zoster heralds the onset of an underlying malignancy [6].

Clinical Picture

Herpes zoster presents in about 80% with a prodrome consisting of hyperaesthesia, paraesthesia, burning disaesthesia, sometimes itching along the affected dermatome (figs. 1, 2). In rare cases these symptoms may be localized to more than one, sometimes several overlapping dermatomes. In general the prodrome lasts only 1–5 days. Rarely, it may precede the appearance of skin lesions by up to 3 weeks. During the prodromal phase herpes zoster may be misinterpreted as cardiac disease, herniated nucleus pulposus or as various gastrointestinal disorders such as cholecystitis, biliary colic, renal colic, appendicitis or even as a gynecologic disorder. A rare condition seen in patients having prodromal symptoms and later dermatome-dependent pain without developing the characteristic rash is called zoster sine herpete [7].

Cutaneous Manifestations

The characteristic zoster rash is unilateral and crossing of the bodies midline is a extremely rare condition, called zoster duplex. Unusually several skin segments may be affected asymmetrically on both sites of the body [8, 9]. As a rule discrete pale to erythematous spots start to develop into a painful asymmetrical unilateral erythema in the affected nerve segment known as 'belt-like pattern' (figs. 2, 3). Normally after a period of about 12–24 h grouped vesicles appear and become confluent (fig. 3). In general, the vesicles are painful and may lead to anxiety and flu-like symptoms in the patients. The subsequent

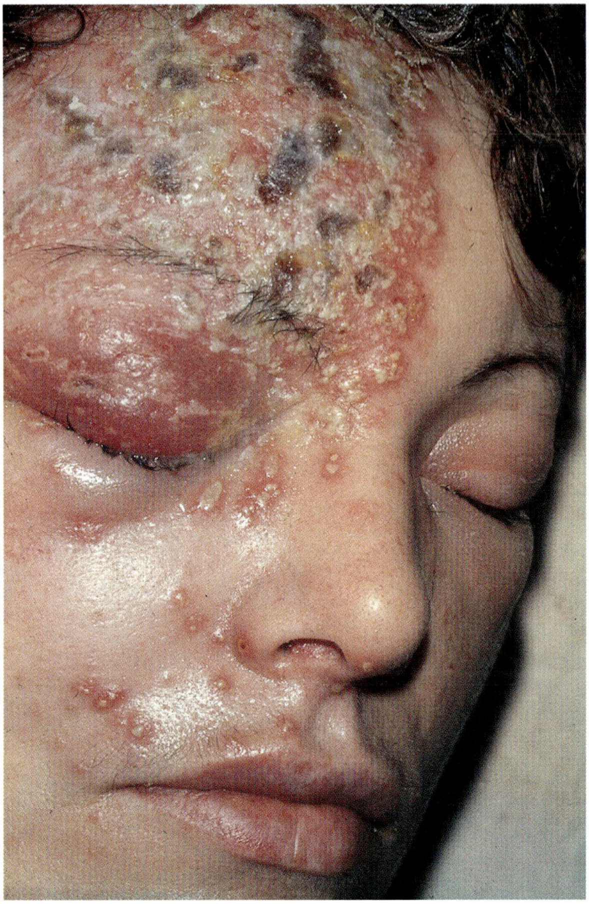

Fig. 1. Herpes zoster of the right ophthalmic nerve (V-1 dermatome) (first branch of the trigeminal nerve) with dissemination of vesicles to V-2-dermatomes (note: severe oedema of the right eye-lid).

stages involve pustulation, erosion, ulceration and crusting (fig. 2). New vesicles may continue to arise during a period of 1–7 days. In immunologically healthy patients duration of rash until disappearance of the crusts takes usually 2–3 weeks. Scarring with hypo- or hyperpigmentation may result [10]. Particularly if the rash is complicated by haemorrhagy and necrosis scarring may be pronounced. In these cases dysaesthesia may develop. The rash is known to be most severe and to last longest in older persons and least severe and shows clearing soonest in children. In immunocompromised patients chronic courses of zosteric skin lesions may last up to several months and development of repeated vesicular and pustular eruptions may be seen [11–13].

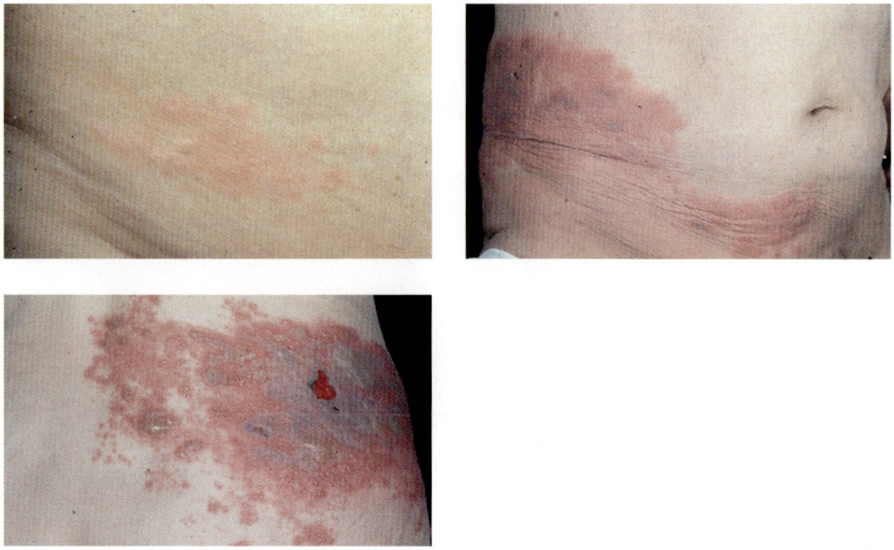

Fig. 2. Early herpes zoster rash: dermatomal erythema at the lower right abdominal skin (TH 11–12 dermatome). Unilateral asymmetrical erythema with initial vesicles limited by the midline of the body (TH 10–12 dermatome). Grouped vesicles and blisters become confluent and erosive.

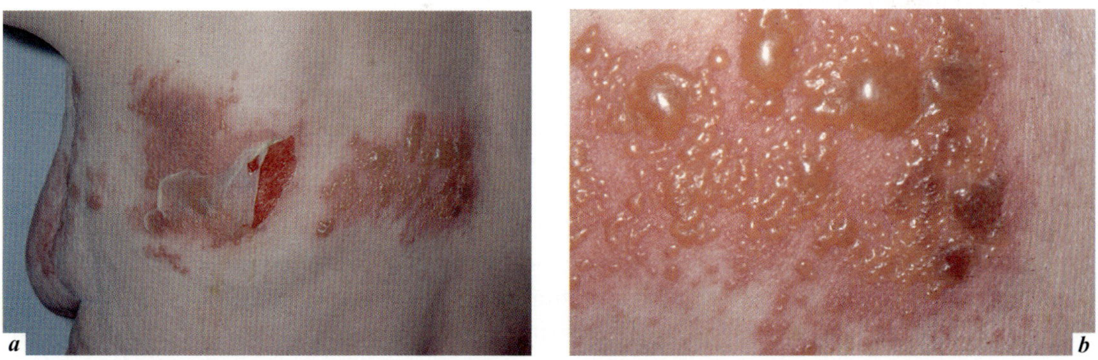

Fig. 3. ***a*** Thoracic Herpes zoster (Th 3–4 segment) on the left side with grouped vesicular and bullous lesions in a 'belt-like pattern'. ***b*** Magnification from fig. 3a.

Localization of Herpes Zoster

Although any dermatome may be involved, dermatomes of the trunk from Th3 to L1 are most frequently affected (figs. 3, 4). The second most frequently involved nerve dermatome is the ophthalmic division of the trigeminal nerve.

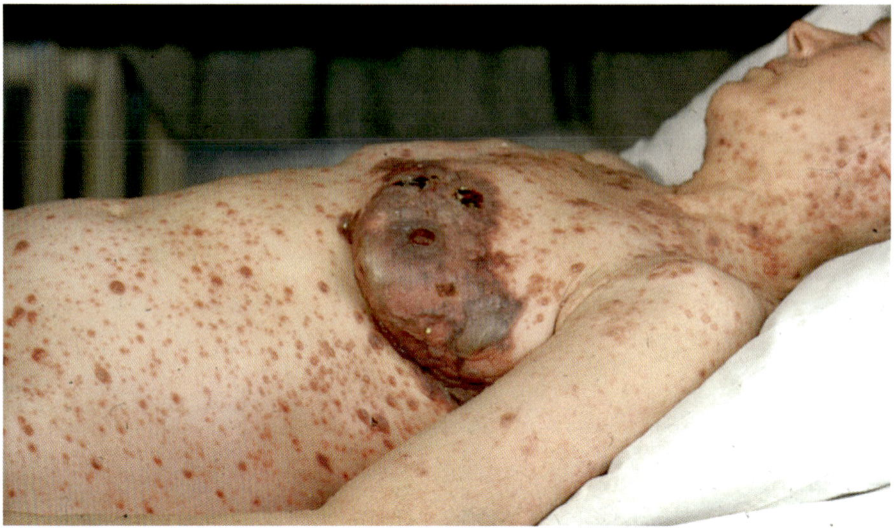

Fig. 4. Disseminated herpes zoster: Thoracic herpes zoster involving the Th 5–6 dermatomes combined with disseminated partially umbilicated vesicles in a 69-year-old woman suffering from Hodgkin's disease.

Herpes zoster of cervical, lumbal and sacral segments is observed in a declining frequency [14]. Haematogenous spread of VZV results in disseminated zoster which is seen in only 1–2% of immunocompetent herpes zoster patients, but which is a particularly more frequent finding in immunodeficient patients [15, 16]. Disseminated herpes zoster (fig. 4) appears in a non-dermatomally pattern of nodules and vesicules and needs sometimes to be differentiated from chickenpox. Involvement of visceral organs such as the lungs and the nervous system is occasionally found in these cases.

According to Meister et al. [14], there is an age-specific predilection of zoster localization. Whereas affection of thoracic segments is preferentially seen in younger patients, the trigeminal nerve (especially the ophthalmic division) and sacral segments are increasingly involved in individuals older than 50 years of age.

Differentiating zoster from other circumscribed rashes may be possible by the characteristic asymmetrical zoster rash and synchronous development of skin lesions starting with erythema followed by vesicular, pustular and finally crustous lesions. Lesions not developing synchronously such as varicella-like lesions, especially reported from AIDS patients, may create a special problem for differential diagnosis [17, 11]. In such cases, virus detection by virus culture, antigen detection by direct immunofluorescence or polymerase chain reaction (PCR) should be made (see chapter: Laboratory confirmation of Herpes Zoster, pp 13–19).

Symptoms

Characteristically the clinical appearance of herpes zoster is accompanied by dermatomal pain, which may be continuous or intermittent and presenting with varying intensity. By definition, pain occurring before and after the dermatomal rash is called zoster-associated-pain. Postzoster neuralgia or PHN is defined as pain, which appears or continues after cutaneous symptoms have disappeared (see chapter: Postherpetic Neuralgia and Other Neurologic Complications, pp 69–80). PHN is the most frequent and important complication of herpes zoster affecting the nervous system [18].

Herpes Zoster Ophthalmicus

Herpes zoster ophthalmicus involves the ophthalmic branch (V-1 dermatome), which is the first division of the trigeminal nerve. According to several studies about 7–18% of reported herpes zoster cases affect the ophthalmic division of the trigeminal nerve [19–21]. While cases of zoster ophthalmicus occur approximately in 10% of zoster patients under the age of 10 years, almost 30% of 80-year-old and older patients suffer from this condition. There is no doubt that ophthalmic zoster is seen particularly more frequently in patients older than 50 years of age [14, 22, 23]. The rash of ophthalmic zoster may extend from the level of the eye to the vertex of the skull. Characteristically it does not cross the midline of the forehead. About 50% of patients with herpes zoster ophthalmicus will develop ocular complications if they do not receive antiviral therapy [12]. Involvement of the nasociliary branch of the ophthalmic nerve which is evidenced by a zosteric rash on the tip and side of the nose (Hutchinson's sign) is seen in about one-third of patients and is usually accompanied by ocular symptoms (fig. 5).

Thus, when ophthalmic zoster affects the side and the tip of the nose, careful attention must be given to the condition of the eye and immediate ophthalmologic consultation is necessary in order to prevent complications of the eye and central nerve system (see chapter: Ophthalmic Manifestations of Herpes Zoster Infection, pp 37–46) [24].

Herpes Zoster Oticus

Herpes zoster affecting the second and third division of the trigeminal nerve and other cranial nerves is likely uncommon. When it occurs, however it may produce symptoms and lesions in the mouth, ear, pharynx or larynx. The Ramsay

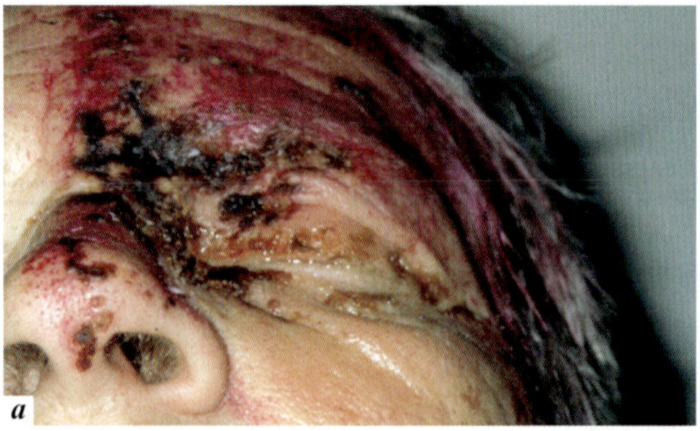

a

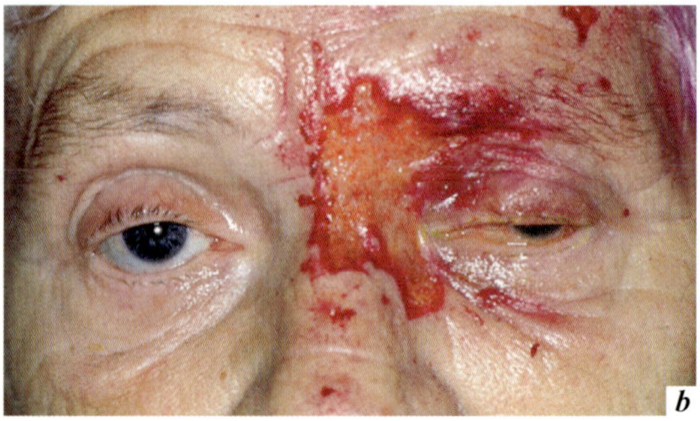

b

Fig. 5. ***a*** Herpes zoster ophthalmicus with involvement of the left tip and side of the nose (nasociliary branch of V-1 (Hutchinson's sign)) in a 84-year-old woman. ***b*** Severe ulceration despite optimal intravenous acyclovir therapy.

Hunt syndrome consists of facial palsy combined with ipsilateral herpes zoster of the external auditory meatus or tympanic membrane. Severe facial pain, facial palsy, decreased hearing or even deafness, tinnitus, vertigo and loss of taste in the anterior part of the tongue may coexist resulting from the involvement of the geniculate ganglion (compare chapter: (Herpes) Zoster Oticus, pp 47–57).

Herpes Zoster in Children

Herpes zoster can develop in immunocompetent children as young as a few weeks of age and should be considered in the differential diagnosis of vesicular eruptions in infants. Zoster in children most frequently involves the thoracic dermatomes (65%). In contrast to adult zoster patients, cranial sites are rarely affected (13 vs. 5%, respectively). Sacral and disseminated zoster are very rare findings (<5%) [25–27].

Herpes zoster certainly occurs in pediatric populations, although by far less frequent compared with adults [28]. Affected children are almost exclusively those, who had an intrauterine exposure to VZV, VZV exposure until the fourth year of age or who are immunosuppressed [25] (fig. 6). Pediatric malignancies most frequently associated with zoster are leukaemia and Hodgkin's lymphoma. As a rule zoster does not precede the clinical manifestations of these malignancies but is associated with chemotherapy or relapse [29]. If immunocompetent children are affected, the course of the disease is rather mild and pain and PHN are extremely uncommon findings. It is quite likely that the vesicular lesions of herpes zoster in this age group are misdiagnosed as impetigo or other cutaneous disorders [30].

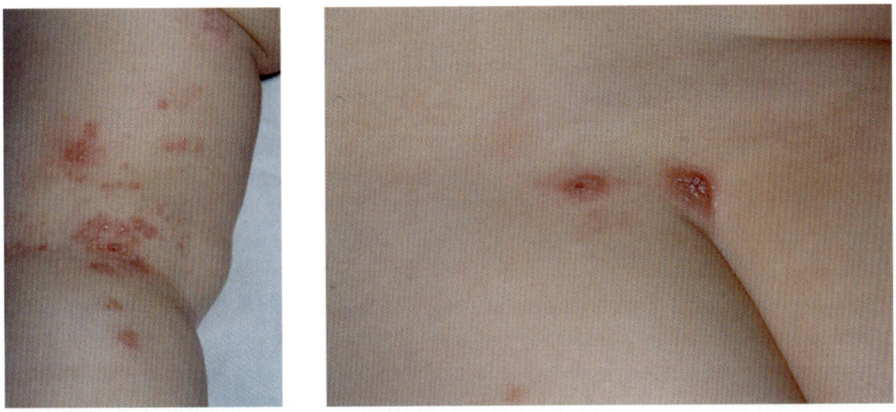

a b

Fig. 6. ***a*** Herpes zoster affecting dermatomes S1–S2 in a 10-months-old girl. ***b*** Grouped vesicles on the left S2 dermatome.

With the increasingly widespread use of varicella-vaccine the frequency of herpes zoster in children will probably continue to decrease. Studies of herpes zoster in normal children [26] have not shown underlying problems and have resulted in the conclusion that laboratory studies in healthy children with herpes zoster are not helpful. The majority of children presenting with herpes zoster is not infected with HIV. Undocumented HIV infection in a young child should however be considered when herpes zoster develops with a short interval after an eruption of chickenpox. This is particularly true if there are no data about the HIV serologic status of the mother [12].

Herpes Zoster in Immunodeficient Patients

Individuals with pathologic or iatrogenic immunodeficiency may present atypical zoster. In some cases, the rash may appear mitigated with only few symptoms. More often the rash shows severe inflammation in part with haemorrhages and necrosis of the skin. Involvement of more than one dermatome may be seen and the rash may persist for a longer period than in immunocompetent persons. Disseminated zoster with varicella-like lesions (fig. 4) combined with involvement of inner organs and the nervous system is more common in immunodeficient persons [16, 31] than in immunocompetent individuals. In persons suffering from lymphocytic leukaemia or Hodgkins disease zoster rash with atypical varicelliform or papular and even verrucous lesions was reported [11, 17]. Since herpes zoster is regarded as an early marker for HIV-infection [32], it is generally accepted to serologically test zoster patients under the age of 50 for HIV infection (see chapter: Herpes Zoster in the Immunocompromised Host, pp 93–106).

Herpes Zoster during Pregnancy

Maternal zoster is hardly considered a risk for the unborn. This is in contrast to maternal varicella (see chapter: Varicella-Zoster Virus Infections during Pregnancy, pp 81–92). Zoster during pregnancy should only be treated with intravenous acyclovir in exceptional cases. In general local and symptomatic treatment is sufficient.

Pregnant zoster patients in a gynaecologic practice, however should be treated in such a way that non-pregnant women without immune protection are not put at risk of varicella [13].

Clinical Diagnosis

The clinical picture of herpes zoster is almost always distinctive enough for diagnosis and laboratory diagnosis is not required. Characteristically, the asymmetrical rash with grouped vesicles is located on inflamed and erythematous dermatomal skin. This and limitation of the rash by the midline as well as prodromal and zoster-associated segmental pain are mostly sufficient to establish the diagnosis. Diagnostic difficulties may arise during the prodromal period and the early phase of zoster and also when atypical skin lesions appear or an atypical site of the body is affected. In such situations laboratory confirmation is required. Equally laboratorial diagnosis is important for VZV infections of pregnant women, newborn infants and particularly if a VZV-infection of the central nervous system is suspected.

The golden standard of laboratorial diagnosis is viral culture (see chapter: Laboratory Confirmation of Herpes Zoster, pp 13–19). The PCR is a useful method to detect VZV-DNA in fluids and in tissues. In case of neurological manifestations PCR is the test of choice to detect VZV in cerebro-spinal fluid [18]. Detection of VZV in blood by the PCR can be predictive of PHN or even diagnostic in some cases of zoster sine herpete. In everyday practice, direct immunofluorescence assays using labelled VZV-specific monoclonal antibodies are a diagnostic approach suitable in terms of low costs and the option to distinguish VZV-infection from HSV-infection. The Tzanck test can be of help to examine the cytopathic effect of VZV in the epidermis with multinucleated giant cells and intranuclear inclusions (fig. 7). The test is done by taking a swab from the base of a blister, transferring the cells to a microscopic glass, and observing the cells under the light-microscope after giemsa staining. This

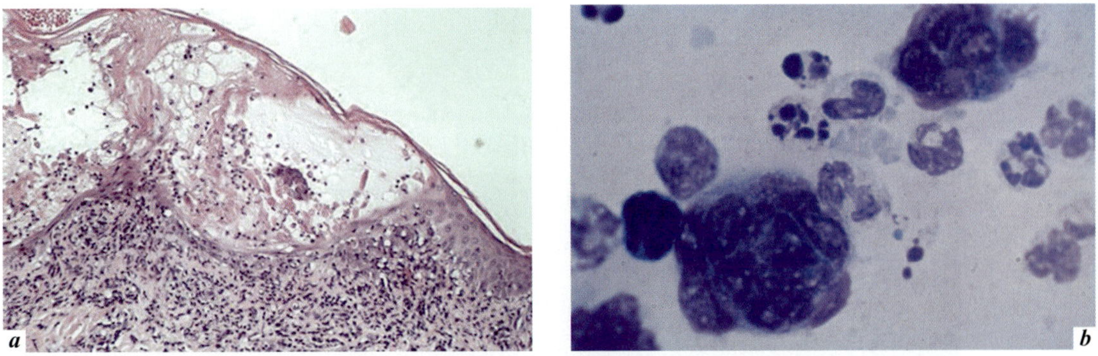

Fig. 7. Histology of a zoster vesicle: subcorneal vesicle with multinuclear giant squamous cell. Tzanck-test: multinuclear giant cells. Features of a herpes virus infection.

simple assay can be done easily at bed-side and may be helpful to establish the diagnosis of herpes virus infection of the skin quickly. Unfortunately, the Tzanck-test is unable to distinguish between VZV- and HSV-infection. Diagnosis of zoster is exceptionally done by histopathological means. The histopathological features with focal intercellular oedema, ballooning degeneration of keratinocytes and multinucleated giant cells as well as intranuclear inclusions (fig. 7) are not characteristic enough to differentiate VZV-infection (herpes zoster and varicella) from HSV-infection.

Serologic tests can provide a retrospective diagnosis when acute and convalescent sera are available. The detection of IgM- and high-titered IgA-anti VZV-antibodies are of some help since they indicate VZV-reactivation whether lesions are present or not [33] (see chapter: Laboratory Confirmation of Herpes Zoster, pp 13–19). In the future detection of lowered VZV-specific cellular immunity might be a very early marker for reactivation of latent VZV-infection indicating zoster eruption.

Differential Diagnosis

Particularly in the prodromal period, herpes zoster is easily confused with other causes of pain. Herpes simplex lesions may be reminiscent of herpes zoster and therefore it may be very difficult to differentiate clinically [12] (figs. 8, 9). The same holds true for bullous or haemorragic erysipelas (fig. 10), contact dermatitis, pannicalitis and insect bites [10]. Autoimmune bullous dermatoses such as bullous pemphigoid or pemphigus may pose special differential diagnostic problems which have to be resolved by specific immunofluorescence tests.

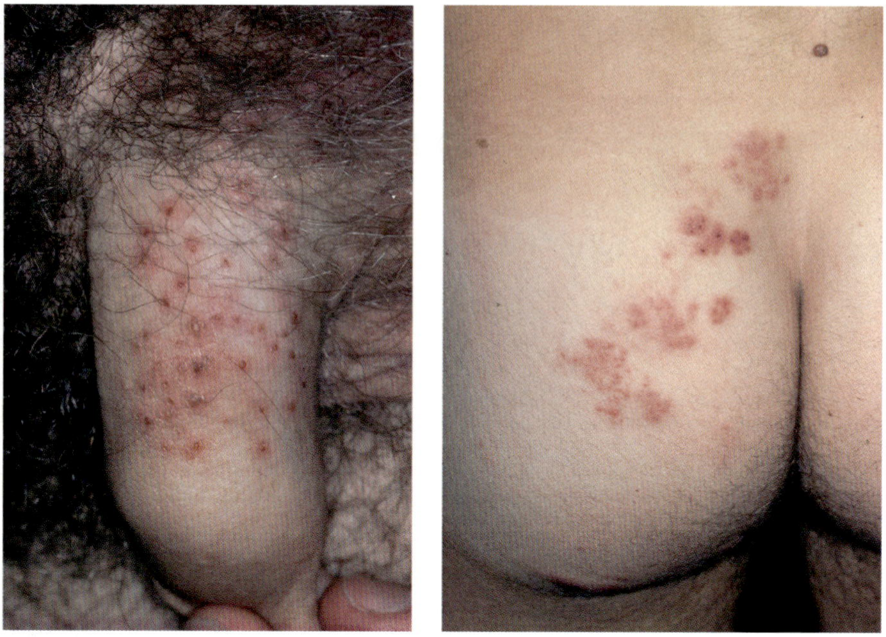

a b

Fig. 8. ***a*** 23-year-old male with vesicular and crusted zoster lesions on the penile shaft (S2 dermatome) reminiscent of genital herpes. ***b*** Asymmetrical grouped erythematous and crusted lesions on the left buttock (S2 dermatome) indicating herpes zoster (same patient).

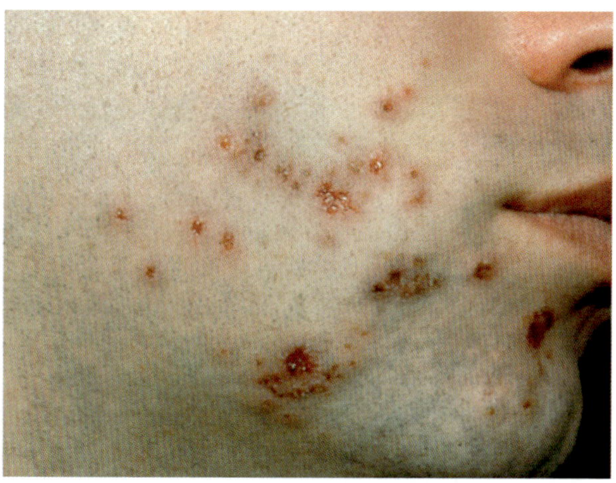

Fig. 9. Zosteriform herpes simplex of the right cheek of a young adult male (no dermatome-associated lesions).

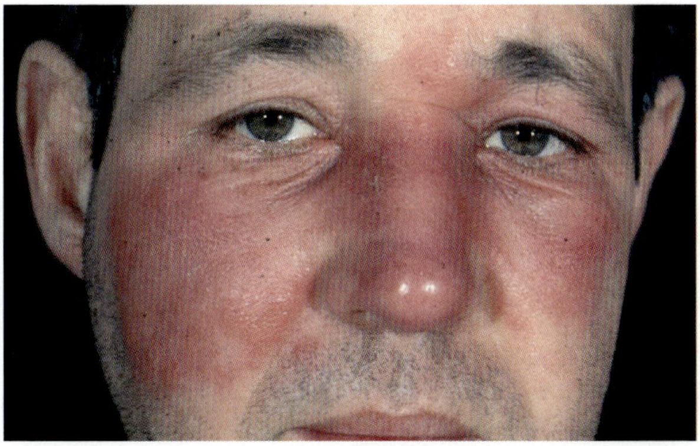

Fig. 10. Facial erysipelas: non-dermatomal patchy infiltrated erythema (note: no limitation of the lesions by the midline).

Complications

While acute and chronic zoster complications affecting skin, eye and central nervous system are quite frequent, complications of visceral organs are very rare findings. At cutaneous and mucocutaneous sites additional secondary bacterial infections may lead to ecthymiform ulcerations and delayed healing which finally may result in either hyper- or hypopigmented scarring (fig. 11a; table 1). Manifestation of psoriasis within the affected dermatomes and also granulomatous reactions may appear [34]. The most important and most frequent complication of shingles is acute pain and particularly chronic pain, also known as PHN. The latter is defined as pain persisting more than 12 weeks after rash healing. Elderly and immunocompromised herpes zoster patients of any age are at increased risk for PHN and also for complications of the central nervous system such as myelitis, encephalitis and motor neuropathy (fig. 11; table 1), (see chapter: Postherpetic Neuralgia and Other Neurologic Complications, pp 69–80). The latter can occur with or without cutaneous manifestations. VZV-infection of cerebral arteries can present as unifocal or multifocal vasculopathy. While unifocal large-vessel vasculopathy (granulomatous arteritis) usually affects elderly immunocompetent persons, multifocal vasculopathy occurs primarily in immunocompromised persons [35, 36]. PHN shows a clear age-dependency. In herpes zoster patients over 70 years its incidence may be as high as 73%. Patients with zoster ophthalmicus and women seem to have a certain increased risk to develop PHN [14, 23]. Immunodeficiency alone, however appears to be no risk for development of chronic pain [25, 37]. Chronic zoster pain in children being immunocompetent or

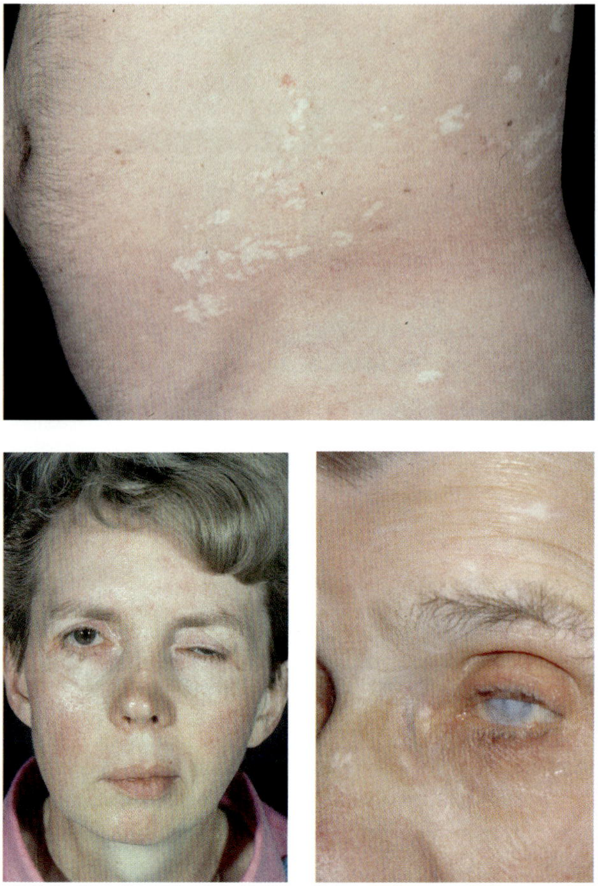

a *b* *c*

Fig. 11. Complications of herpes zoster: ***a*** Hypopigmented postzosteric scars on the left lower abdominal skin (Th 10 and 12 dermatomes). ***b*** Ramsay-Hunt Syndrome with paralysis of facial muscles (Bell's palsy). ***c*** 75-year-old woman: Blindness of the left eye with zosteric scars on the left forehead and nose (V-1 dermatome).

immunosuppressed is extremely rare [38]. Ocular complications may present as zoster lesions of the eyelid with subsequent scar formation, conjunctivitis, corneal erosions, episcleritis, iritis, stromal keratitis and ulcers (see chapter: Ophthalmic Manifestations of Herpes Zoster Infections, pp 37–46). A severe complication with visual impairment is acute retinal necrosis, which has been described both in immunocompetent and in immunocompromised patients. Finally it may lead to blindness (fig. 11c). This condition is regularly preceded by zoster ophthalmicus. However, it may also develop in combination with

Table 1. Complications of zoster

	Cutaneous and mucocutaneous sites	Nervous system	Eye	Inner organs
Acute complications	Bacterial secondary infections	Encephalitis	Conjunctivitis	Pneumonia
	Zoster haemorrhagicus	Meningitis	Episcleritis/scleritis	Esophagitis
	Zoster gangraenosus	Granulomatous arteritis	Uveitis	Myocarditis
	Zoster generalisatus	Segmental pareses	Keratitis	Enterocolitis
		Facial paresis in zoster oticus	Iridocyclitis (→ glaucoma)	Pancreatitis
Chronic complications	Persisting zoster	PHN	Keratitis	Arthritis
	Scar formation (atrophic scars, hypertrophic scars)	Guillain-Barré syndrome	Chorioretinitis	
	Hypo/depigmentation	Myelitis	Retrobulbar neuritis	
	Granulomatous skin lesions	Motor neuropathy	Vasculitis	
	Pseudolymphoma	Abdominal hernias	Panophthalmitis	
	Manifestation of psoriasis (Köbner's phenomenon)	Phrenoplegia	Atrophy of optic nerve	
		Bladder dysfunction		

herpes zoster in a remote dermatome. Acute retinal necrosis can be stopped successfully with early and consequent intravenous antiviral therapy, as shown in clinical case reports [39] (see chapter: Ophthalmic Manifestations of Herpes Zoster Infection, pp 37–46). In HIV infected patients, however, such necrotic lesions respond only partially to systemic i.v. acyclovir therapy with the risk of consecutive blindness [24].

Psychosocial Issues

The course of herpes zoster during childhood is generally benign and almost always there is no progressive pain or discomfort. Contrary to adult herpes zoster patients complications and sequelae are quite common. The total duration of the disease is generally between 10 and 14 days. Nevertheless, the course may take as long as 3–4 weeks before the skin returns to normal. The most debilitating complication of herpes zoster is acute neuritis and PHN. This accounts both for immunocompromised and for normal herpes zoster patients.

While PHN is extremely uncommon in young individuals, at least every second herpes zoster patient over the age of 60 will have persistent and piercing pain. The average duration of PHN is about 6 months in most cases. In the 6 months after onset of herpes zoster up to 10% of patients return to the physician with recurrent complaints [18]. This complication has led to a variety of treatment regimens such as long-term administration of analgesics and very early administration of systemic antiviral therapy together with analgesics in acute herpes zoster (see chapter: Postherpetic Neuralgia and Other Neurologic Complications, pp 69–80). It has become clear that herpes zoster may have a tremendous impact on the everyday's life of affected individuals older than 50 years of age. The patients may suffer from fluctuating levels of pain for many months or even years. As a consequence treatment of PHN can lead to a range of drug therapies. Nevertheless, the chronicity and intensity of chronic pain can lead to depression and even suicide, particularly among patients over 70 years of age. In older patients who experience the disease on the head and face, which are approximately 15% of all zoster patients, a slight touch, contact with clothing or even a breeze can trigger an unbearable pain (allodynia) [18]. During or after shingles focal itch also may develop. So far this complaint has attracted only little medical attention. According to Oaklander et al. [40] postherpetic itch can lead to self-injury from scratching skin that has lost protective sensation. Postherpetic itch seems to occur particularly in patients with herpes zoster involving the sensory innervation region of the trigeminal nerve or other cranial nerves. Physicans should be aware of the importance to treat shingles early and aggressively with systemic antivirals, analgesics and tricyclic antidepressants to minimize zoster-related complications. Urgent indications for systemic antiviral therapy are herpes zoster in patients beyond the age of 50 years and patients at any age with herpes zoster in the head and neck area. Another urgent indication is severe herpes zoster on the trunk and on the extremities, herpes zoster in immunosuppressed patients and zoster in patients with severe atopic dermatitis and severe eczema [13, 41]. Once PHN has established, an interdisciplinary approach to the care of these patients should comprise behavioural and physical therapy, as well as supportive psychotherapy [40].

References

1 Balfour HH: Varicella zoster virus infections in immunocompromised hosts. A review of the natural history and management. Am J Med 1988;85:68–73.

2 Hope-Simpson RE: The nature of herpes zoster: a long term study and new hypothesis. Proc R Soc Med 1965;58:9–20.

3 Friedman-Kien AE, Lafleur FL, Gendler E: Herpes zoster: a possible early clinical sign for development of acquired immunodeficiency syndrome in high-risk individuals. J Am Acad Dermatol 1986;14:1023–1028.

4 Kost RG, Straus SE: Postherpetic neuralgia – pathogenesis, treatment, and prevention. N Engl J Med 1996;335:32–42.

5 Aliegro MB, Dorruccci M, Pezzotti P, Rezza G, Sinicco A, Barbanera M: Herpes zoster and progression to AIDS in a cohort of individuals who seroconverted to human immunodeficiency virus. Italian HIV Seroconversion Study. Clin Infect Dis 1996;23:990–995.

6 Smith JB, Fenske NA: Herpes zoster and internal malignancy. South Med J 1995;88:1089–1092.
7 Rudra T: Zoster sine herpete. Br J Clin Pract 1990;44:284.
8 Vu AQ, Radonich M, Heald PW: Herpes zoster in seven desperate dermatomes (zoster multiplex): report of a case and review of the literature. J Am Acad Dermatol 1999;40:868–869.
9 Bloss G, Ebisch MA, Kunz M, Gross G: Bilateral asymmetric herpes zoster in adolescence. Hautarzt 2001;52:335–338.
10 Gross G: Herpes zoster. Dtsch Med Wochenschr 1997;122:132–139.
11 Hoppenjans WB, Bibler MR, Orne RL: Prolonged cutaneous herpes zoster in acquired immunodeficiency syndrome. Arch Dermatol 1990;126:1048–1050.
12 Gnann JW Jr, Whitley RJ: Clinical practice. Herpes zoster. N Engl J Med 2002;347:340–346.
13 Gross G, Schöfer H, Wassilew S, Friese K, Timm A, Guthoff R, Pau HW, Malin JP, Doerr HW: Herpes zoster guideline of the German Dermatology Society (DDG). J Clin Virol 2003;26: 277–289.
14 Meister W, Neiss A, Gross G, Doerr H, Holbel W, Malin J, von Essen J, Reimann B, Witke C, Wutzler P: Demography, symptomatology, and course of disease in ambulatory zoster patients. A physician-based survey in Germany. Intervirology 1998;41:272–277.
15 Meier JL, Strauss SE: Comparative biology of latent varicella-zoster virus and herpes simplex virus infections. J Infect Dis 1992;166(suppl 1):S13–S23.
16 Cohen RR, Beltraini VP, Grossmann ME: Disseminated herpes zoster in patients with human immunodeficiency virus infection. Am J Med 1998;84:1076–1080.
17 Gallagher JG, Merigan TC: Prolonged herpes-zoster infection associated with immunosuppressive therapy. Am Intern Med 1979;91:842–846.
18 Gilden DH, Kleinschmidt-DeMasters BK, La Guardia JJ, Mahalingam R, Cohrs RJ: Neurological complications of the reactivation of varicella-zoster virus. N Engl J Med 2000;342:635–645.
19 Burgoon CF, Burgoon JS, Baldrige GD: The natural history of herpes zoster. JAMA 1957;164: 265–269.
20 De Moragas JM, Kierland R: The outcome of patients with herpes zoster. Arch Dermatol 1957;75:193–196.
21 Ragozzino MW, Melton LJ III, Kurland LT, Chu CP, Perry HO: Population-based study of herpes zoster and its sequelae. Medicine (Baltimore) 1982;61:310–316.
22 Hardening SP, Lipton JR, Wells JCD: Natural history of herpes zoster ophthalmicus: predictors of postherpetic neuralgia and ocular involvement. Br J Ophthalmol 1987;71:353–358.
23 Meister W, Neiss A, Gross G, Doerr HW, Holbel W, Malin JP, von Essen J, Reimann BY, Witke C, Wutzler P: A prognostic score for postherpetic neuralgia in ambulatory patients. Infection 1998;26:359–363.
24 Liesegang TJ: Diagnosis and therapy of herpes zoster ophthalmicus. Ophthalmology 1991;98: 1216–1229.
25 Guess HA, Broughton DD, Melton LJ III, Kurland LT, Chu CP, Perry HO: Epidemiology of herpes zoster in children and adolescents: a population-based study. Pediatrics 1985;76:512–517.
26 Smith CG, Glaser DA: Herpes zoster in childhood: case report and review of the literature. Pediatr Dermatol 1996;13:226–229.
27 Terada K, Kawano S, Yoshihiro K: Characteristics of herpes zoster in otherwise normal children. Pediatr Infect Dis J 1993;12:960–961.
28 Donahue JG, Choo PW, Manson JE: The incidence of herpes zoster. Arch Intern Med 1995;155:1605–1609.
29 Bacon GE, Olivier WJ, Shapiro BA: Factors contributing to severity of herpes zoster in children. J Pediatr 1965;67:768–771.
30 Kurlan JG, Conelly BL, Lucky AW: Herpes zoster in the first year of life following postnatal exposure to Varicella-zoster virus. Arch Dermatol 2004;140:1268–1272.
31 Rusthoven JJ, Ahlgren P, Elhakim T, Pinfold P, Reid J, Stewart L, Feld R: Varicella zoster infection in adult cancer patients: a population study. Arch Intern Med 1988;148:1561–1566.
32 Melbye M, Grossman RJ, Goebert JJ: Risk of AIDS after herpes zoster. Lancet 1978;1:728–730.
33 Doerr HW, Rentschler M, Scheifler G: Serologic detection of active infections with human herpesviruses (CMV, EBV, HSV, VZV): diagnostic potential of IgA class and IgG-subclass-specific antibodies. Infection 1987;15:93–98.

34 Requena L, Kutzner H, Pescalonilla P, Otiz S, Schaller J, Rohwedder A: Cutaneous reactions at sites of herpes zoster scars: an expanded spectrum. Br J Dermatol 1998;138:161–168.
35 Gilden DH: Varicella zoster virus vasculopathy and disseminated encephalomyelitis. J Neurol Sci 2002;195:99–101.
36 Gilden DH, Lipton HL, Wolf JS, Akenbrandt W, Smith JE, Mahalingam R, Forghani B: Two patients with unusual forms of varicella-zoster virus vasculopathy. N Engl J Med 2002;347:1500–1503.
37 Hope-Simpson RE: Postherpetic neuralgia. J R Coll Gen Pract 1975;25:571–575.
38 Malin JP: Zoster and the nervous system. Dtsch Med Wochenschr 1996;121:635–638.
39 Nau R, Lansch M, Stiefel M, Polak T, Reiber H: Varicella zoster virus-associated focal vasculitis without herpes zoster: recovery after treatment with acyclovir. Neurology 1998;51:914–915.
40 Oaklander AL, Bowsher D, Galer B, Haanpaa M, Jensen P: Herpes zoster itch: preliminary epidemiologic data. J Pain 2003;4:338–343.
41 Volpi A, Gross G, Hercogova J, Johnson RW: Current management of herpes zoster: the European view. Am J Clin Dermatol 2005;6:317–325.

Gerd Gross, MD
Professor of Dermatology and Venereology, Director
Department of Dermatology and Venereology, University of Rostock
Augustenstrasse 80–84
DE–18055 Rostock (Germany)
Tel. +49 381 4949700, Fax +49 381 4949702, E-Mail gerd.gross@med.uni-rostock.de

Gross G , Doerr HW (eds): Herpes Zoster.
Monogr Virol. Basel, Karger, 2006, vol 26, pp 37–46

Ophthalmic Manifestations of Herpes Zoster Infection

Hermann Guembel

Bundeswehrkrankenhaus Ulm, Academic Teaching Hospital of the University Clinics of Ulm, Ulm, Germany

The varicella–zoster virus (VZV) frequently reactivates within the ophthalmic part of the trigeminal nerve [1, 2]. Herpes zoster ophthalmicus (HZO) is manifested by localized pain over the distribution of the first division of the trigeminus cutaneous vesicular eruption around the forehead. There are also frequently lymph node enlargement in the defined area, headache, sometimes neck stiffness and a red eye.

Approximately 10–25% [3] of all cases of herpes zoster affect the ophthalmic division of the trigeminal nerve. The condition is then referred to as HZO.

Direct ocular involvement is specifically correlated with age, gender, or severity of disease. Serious sequelae include chronic inflammation, vision loss and disabling pain. HZO starts normally with an influenza-like illness with malaise and low-grade fever that lasts up to 5 days before the rash on the forehead appears. Subsequently, erythematous macules appear along the involved dermatome, rapidly progressing over several days to papules and vesicles containing clear serous fluid and pustules. These lesions rupture and get crusted. Very rarely, the eye can become involved when the maxillary nerve is involved. Affection of the nasociliary nerve (Hutchinson's sign) is quite common and supplies the side and the tip of the nose. The Hutchinson's sign occurs in about one-third of patients with HZO and correlates significantly with ocular complications [4].

However, there is no correlation between ocular complications and age, sex or severity of the skin rash. Severe ocular complications may therefore develop in patients with only a slight rash. Approximately 50% of people with VZV infection involving the ophthalmic division of the fifth cranial nerve have ocular involvement, often in association with typical skin lesions.

Immunocompromised persons, particularly those with human immunodeficiency virus (HIV) infection, have a much higher risk of developing HZO than the normal population [5]. Viral transmission from patients with HZO can occur, but is less frequent than transmission from patients with chickenpox [6].

The risk of ocular complications and subsequent visual loss in ophthalmic zoster disease, determined by individually unique virus–host interactions and the efficacy of antiviral treatment, are still uncertain.

Pathophysiology

It is the reactivation of latent VZV within the first division of the trigeminal ganglion which may lead to a wide range of ocular inflammatory disorders, particularly when the nasociliary branch is affected.

After primary VZV infection leads to chickenpox nucleic acids of VZV remain dormant in sensory ganglia. There is following a phase of latency, which usually lasts for several decades before VZV reactivation occurs. The duration of VZV-DNA shedding from the affected sensory nerve endings within the ocular epithelium in HZO has not yet been studied extensively. In some cases of acute ophthalmic zoster disease VZV-DNA was detected in conjunctival swabs [7]. Another study showed that the duration of VZV-DNA shedding in HZO was highly variable and age dependent. Elderly people, of 66 years or older, had a significantly more prolonged duration of VZV-DNA presence in the conjunctival swabs than patients who were younger than 66 years [8].

HZO represents reactivated VZV that travels down the ophthalmic nerve ganglion, taking 3–4 days to reach the nerve endings. Because the nasociliary branch innervates both the tip and the homolateral side of the nose as well as the cornea, most serious ocular involvement will develop if this branch is affected. A classic sign is the Hutchinson's sign, which has been a predictor of affected ocular structures. Ocular (table 1) and cranial nerve (table 2) involvement is presented with the time after onset of rash.

Stages of Infection

Acute Phase

The rash may be maculo-verricular but also papular with pain of varying degree at the dermatome related to the ophthalmic nerve. In figure 1, a 28-year-old male patient with a presumed HIV infection is shown who developed a severe erythematous and rash. The lesions varied in size, thickness and both were scattered and confluent, partially hemorrhagic.

Table 1. Herpes zoster ophthalmicus – disorders of the anterior eye segment

Involved structures	Signs	Starting clinical symptoms (day 0 = onset of, up to weeks)
Eyelid and conjunctiva		
Blepharoconjunctivitis	cutaneus rash respecting midline	day 0 and painful dermatom
	conjunctival edema/inflammation	days 2–3
	vesicular crusts and lesions	6 days
Secondary bacterial infection	yellow colored crusts	1–2 weeks
Episclera/sclera		
Episcleritis/scleritis	diffuse or localized redness, edema and pain	5–8 days
Cornea		
Punctate epithelial keratitis	edema of corneal surface	1–2 days
Dendritic keratitis	worm-like epithelial lesions	4–6 days
Anterior stromal keratitis	infiltrating stromal structures	1–2 weeks
Deep stromal keratitis	lipid infiltrates and deep stromal inflammation and neovascularization	1 month to years
Neurotrophic keratopathy	punctate keratitis and erosions persistent epithelial defects	months to years
Anterior chamber		
Uveitis	inflammation, cells and synechias	2 weeks to years

Table 2. Herpes zoster ophthalmicus – disorders of the posterior eye segment

Involved structures	Signs	Starting clinical symptoms (day 0 = onset of, up to weeks)
Retina		
Acute retinal necrosis syndrome/ progressive outer retinal necrosis syndrome	retinal whitening, intraretinal hemorrhages occlusive vasculitis, vitreous floaters	no time schedule possible
Cranial nerves		
Neuritis nervi optici	edema of the optic nerve	no time schedule possible
Oculomotor palsies	motional abnormalities of the eye muscles	no time schedule possible

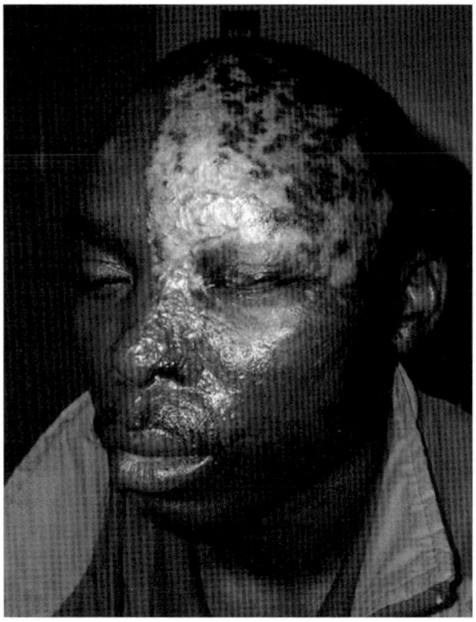

Fig. 1. Herpes zoster ophthalmicus skin manifestations of a immunocompromised patient (presumed HIV+) strictly obey the midline involving supraorbital, lacrimal and nasociliary and maxillary branches of the trigeminal nerve.

Ocular Involvement

Eyelids are commonly swollen and red. Ptosis can be observed in most patients. Some develop blepharitis and the vesicular lesions mostly resolve with scarring.

Conjunctivitis is seen very often after onset of rash. Mostly there is edema and inflammation as well as petechial hemorrhages. The conjunctival symptoms resolve within 1 week, but can be prolonged by secondary bacterial superinfection (e.g. *Staphylococcus aureus*).

Episcleritis and scleritis (fig. 2) are not rare findings and require a frequent slitlamp control. It might involve the cornea if it is present for longer than one week.

Keratitis is very common and presents in several forms.

The epithelial form of keratitis is seen frequently within the first 2 days after onset of rash. There are small, fine precipitates of the cornea resolving spontaneously a few days later. Diagnosis is done by staining with flourescein or bengalrosa.

Nummular keratitis and disciform keratitis (fig. 3) are not very common and are detected about 10–21 days after onset of rash. Fine granular deposits or disciform lesions can be localized near Bowman membrane. There is also a stromal haze surrounding the lesions.

Anterior uveitis is seen quite frequently and will only be visible with careful slit lamp biomicroscopic examination. It develops within 2 weeks after onset of the rash. As a result of severe inflammation iris atrophy may result. Also endothelial dysfunction of the cornea may occur leading to edema with central vision loss [9].

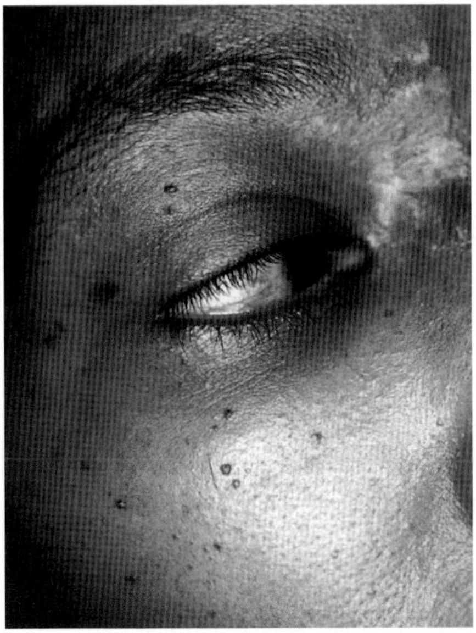

Fig. 2. Residual episcleritis and scarring following zoster ophthalmics involving the nasociliary branch.

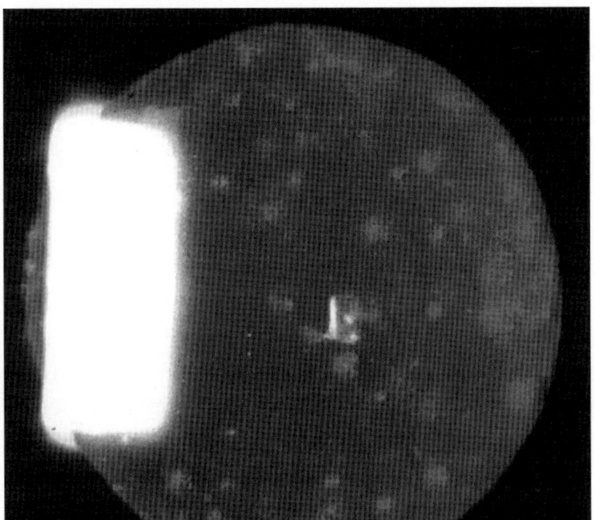

Fig. 3. Nummular keratitis of a patient after HZO with subepithelial infiltrates, located in the anterior stroma by slit lamp.

Chronic Phase

From a clinical point of view there are more cases presenting signs of severe chronical involvement of the sclera and cornea than with acute clinical findings.

Stromal or deep stromal keratitis is a later stage of stromal keratitis characterized by infiltrates of differing degree. It develops 3–4 months after the first onset of inflammation. Usually it is localized in the center of the cornea. A chronic relapse of keratitis is not unusual and as a result of a uncontrolled chronic disease deep corneal neovascularizations and lipid infiltrates may be detectable in time.

Neurotrophic keratopathy is a clinical endpoint of keratitis. Destruction of the cornea may lead to decreased corneal sensitivity, often followed by injuries. Another pathway to vision loss results in corneal thinning with bullous keratopathy and corneal perforation.

Acute retinal necrosis syndrome (ARN) and progressive outer retinal necrosis syndrome (PORN) are fortunately rare findings in young patients and in immunocompromised patients (fig. 4). First signs of ARN and PORN are blurred vision in one or both eyes (30% bilateral involvement) and pain. Clinically the fundus of ARN shows whitening and peripheral patches with occlusive vasculitis and vitreous inflammation. In contrast in PORN vitreous cells are absent, since immunocompromised patients are unable to produce an inflammatory response. In such cases visual prognosis is bad despite antiviral therapy and more than 70% develop retinal detachment.

Differential Diagnosis

Because of the typical clinical findings (rash, pain etc.) in the acute phase of HZO there is no really clear differential diagnosis. Erysipelas may however initiate herpes zoster at an early stage. A chronic HSV infection can imitate similar corneal destruction and thinning as seen in HZO with ocular affection.

Diagnostic Tools in Herpes Zoster Ophthalmicus

There are several aspects to detect VZV in the eye during a chronic course of HZO. It is easy to diagnose acute HZO with its typical rash, involving the nasociliary branch of the ophthalmic nerve. There are quite more difficulties to

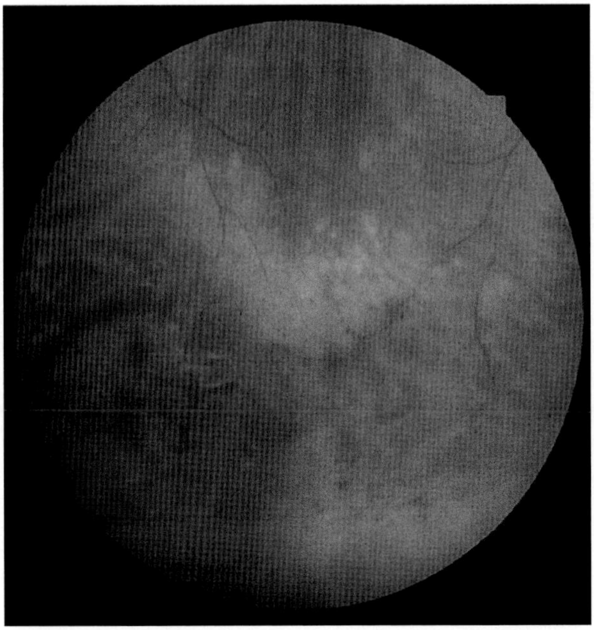

Fig. 4. Acute retinal necrosis is defined with peripheral patches of necrotic zones which rapidly coalesce and may rapidly progress to involve the macula.

find out the etiology of a posterior uveitis or of other intraocular diseases. There is a need of early and prompt diagnosis in patients with viral intraocular infections. Polymerase chain reaction (PCR) is the diagnostic tool in virology, which is sensitive and rapid enough. In addition the PCR technique requires only a small sample. In 11 cases with posterior uveitis Gargiulo et al. [10] found three of five cases suffering from ARN VZV and in the other two cases HSV1–2 by a PCR based assay.

Serological diagnosis by enzyme-linked immunosorbent assay and the immunofluorescence technique is only efficient and helpful in immunodeficient patients suffering from VZV reactivation [11].

Differential Diagnosis

The most common differential diagnoses of HZO are zosteriform herpes simplex and bacterial infections like erysipel. When the clinical presentation is

Table 3. Treatment of HZO infection with ophthalmic involvement based on immune status and age

Immune status	Age	Systemic corticosteroids	Antivirals	Topical therapy	Other medications
Immunodeficient	any	none	i.v. acyclovir* 15 mg/kg/bw/day or valacyclovir 1,000 mg 3 × day	acyclovir ointment in keratitis 5 × day foscarnet 2.4 mg/0.1 ml, i.vitr. in ARN/PORN	optional topical prednison 5 × day and mydriaticum in anterior uveitis
Immunocompetent	over 60 years	40 mg/7 days 20 mg/7 days 10 mg/7 days	oral acyclovir 800 mg/5 × day or valacyclovir 1,000 mg 3 × day or famciclovir 250 mg 3 × day or brivudin 125 mg 1 × day	acyclovir ointment in keratitis 5 × day, foscarnet 2.4 mg	optional topical prednisone 5 × day and mydriaticum in anterior uveitis

i.v. = Intravenous, *the only antiviral which could be applied i.v., i.vitr. = intravitreal application of a drug by injection in the vitreous.

not typical, viral detection should be done by PCR, direct immunofluorescence assay or by serology [12].

Therapy

The guidelines of the German Dermatology Society (DDG) which are based on the experience of the authors and the review of the literature [13] present four systemic antivirals, acyclovir, valacyclovir, famciclovir and brivudin, which are approved for the treatment of herpes zoster infections.

In table 3, the current antiviral therapy for HZO patients is listed in accordance to immune status and age. Especially in patients with ARN or PORN additionally a single intravitreal injection of foscarnet is recommended as soon as possible to stop viral replication and further progression of the retinitis [14]. Concerning treatment of zoster in general, some orally given antivirals have a better potential to lower postherpetic neuralgia (PHN) than others. Brivudin [15] had an 11% lower PHN rate than acyclovir [16] and was as effective as famciclovir to reduce zoster associated pain [17].

It is well known that VZV may become resistant to nucleoside analogs such as the above mentioned antivirals acyclovir, valacyclovir, famciclovir and brivudin. The resistance is mainly due to a mutation of the thymidin kinase. Drug of choice in patients with resistant strains of VZV is intravenous foscarnet, 3 times 40 mg or twice 50 mg per kg body weight per day [18]. In cases where foscarnet would also be ineffective, treatment with cidofovir is indicated as an alternative [19]. Both antivirals have some severe side effects (nephrotoxicity, ocular hypotony etc.) and should be admitted very carefully [18, 19].

References

1 Hope-Simpson RE: The nature of herpes zoster: a long term study and new hypothesis. Proc R Soc Med 1965;58:9–20.

2 Wilson JB: Thirty one years of herpes zoster in a rural practice. Br Med J 1986;22:1349–1351.

3 Naumann G, Gass JD, Font RL: Histopathology of herpes zoster ophthalmicus. Am J Ophthalmol 1968;65:533–541.

4 Liesegang TJ: Herpes zoster virus infection. Curr Opin Ophthalmol 2004;15:531–536.

5 Sandor EV, Millman A, Croxson TS, Mildvan D: Herpes zoster ophthalmicus in patients at risk for the acquired immune deficiency syndrome (AIDS). Am J Ophthalmol 1986;101: 153–157.

6 Baratz KH, Goins K, Cobo M: Varicella-zoster viral infections; in Kaufman HE (ed): The Cornea. New York, Livingstone, 1988.

7 Yamamoto S, Pavan-Langston D, Kinoshita S, Nishida K, Shimomura Y, Tano Y: Detecting herpes virus DNA in uveitis using polymerase chain reaction. Br J Ophthalmol 1996;80:465–468.

8 Zaal MJW, Völker-Dieben HJ, Wienesen M, DÀmaro J, Kijlstra A: Longitudinal analysis of varicella-zoster virus DNA on the ocular surface associated with herpes zoster ophthalmicus. Am J Ophthalmol 2001;131:25–29.

9 Zaal MJW, Völker-Dieben HJ, DÁmaro J: Visual prognosis in immunocompetent patients with herpes zoster ophthalmicus. Acta Ophthalmol Scand 2003;81:216–220.

10 Gargiulo F, Francesco de MA, Nascimbeni G, Turano R, Perandin F, Gandolfo E, Manca N: Polymerase chain reaction as a rapid diagnostic tool for therapy of acute retinal necrosis syndrome. J Med Virol 2003;69:397–400.

11 Wutzler P, Doerr HW: Herpes zoster: a herpes non-simplex disease. Dtsch Med Wochenschr 1995;120:1133–1138.

12 Gross G, Doerr HW: Herpes zoster guidelines of German Dermatological Society. J Clin Virol 2003;27:308–309.

13 Gross G, Schöfer H, Wassilew S, Friese K, Timm A, Guthoff R, Pau HW, Malin JP, Wutzler P, Doerr HW: Herpes zoster guideline of the German Dermatology Society (DDG). J Clin Virol 2003;26:277–289.

14 Gümbel H, Ohrloff C: Opportunistic infections of the eye in immunocompromised patients. Ophthalmologica 1997;211:53–61.

15 Vij O, Bornfeld N, Roggendorf M, Fiedler M, Schilling H: Brivudin as an alternative systemic therapy to acyclovir and ganciclovir in acute retinal necrosis syndrome due to varicella zoster virus. Klin Monatsbl Augenheilkd 2003;220:710–715.

16 Wassilew SW, Wutzler P, Brivudin Herpes Zoster Study Group: Oral brivudin in comparision with acyclovir for herpes zoster: a survey study on postherpetic neuralgia. Antiviral Res 2003;59: 57–60.

17 Wassilew SW, Wutzler P, Brivudin Herpes Zoster Study Group: Oral brivudin in comparison with acyclovir for improved therapy of herpes zoster in immunocompetent patients: results of a randomized double-blind, multicentered study. Antiviral Res 2003;59:49–56.

18 Breton G, Fillet AM, Katlama C, Bricaire F, Caumes E: Acyclovir-resistant herpes zoster in human immunodeficiency virus-infected patients: results of foscarnet therapy. Clin Infect Dis 1998;27: 1525–1527.
19 Safrin S, Cherrington J, Jaffe HS: Clinical uses of cidofovir. Rev Med Virol 1997;7:145–156.

Hermann Guembel, MD
Bundeswehrkrankenhaus Ulm
Academic Teaching Hospital of the University Clinics of Ulm
Oberer Eselsberg 40
DE–89081 Ulm (Germany)
Tel. +49 731 17101401, Fax +49 731 17101407, E-Mail hermann.guembel@extern.uni-ulm.de

Gross G, Doerr HW (eds): Herpes Zoster.
Monogr Virol. Basel, Karger, 2006, vol 26, pp 47–57

(Herpes) Zoster Oticus

H.W. Pau

Department of Otorhinolaryngology, Head and Neck Surgery, Medical Faculty, University of Rostock, Rostock, Germany

In literature, the term '(herpes) zoster oticus' is mostly attributed to the neurologist James Ramsay Hunt, who described the disease in 1907 [1]. To our knowledge, however, he was proceeded by the German otologist Otto Körner (one of the author's predecessors in Rostock), who published about 'zoster oticus' in 1904 [2].

The disease is the clinical expression of an infection with – or better reactivation of – the varicella–zoster virus (VZV) in the head – especially the ear region. The incidence is roughly estimated as 1% of all zoster localizations. Patients with (herpes) zoster oticus typically present with earache, vesicular eruptions in the auricle or its surroundings (sometimes also in the buccal or oropharyngeal mucosa) and different alterations of the VII, VIII (sometimes also V, IX, and X) cranial nerves [3] (fig. 1).

VZV (first contact: chickenpox, varicella) may persist in various, mostly neuronal structures – in case of zoster oticus most often in the geniculate ganglion. From these sites, dormant viruses may be reactivated by certain stimuli, causing the clinical outbreak of the disease [4, 5]. However, occasional cases could be attributed to reinfection of immunized hosts by re-exposition to VZV (contact with varicella or zoster patients) [6].

Etiology and Pathogenesis

In 1907, Ramsay Hunt suggested that the disease resulted from a geniculate ganglionitis after reactivation of surviving viruses (VZV) – a theory that recently proved true for many cases examined [2, 7]. In histological temporal bone sections, VZV DNA could be identified in the region of the geniculate ganglion using the polymerase chain reaction.

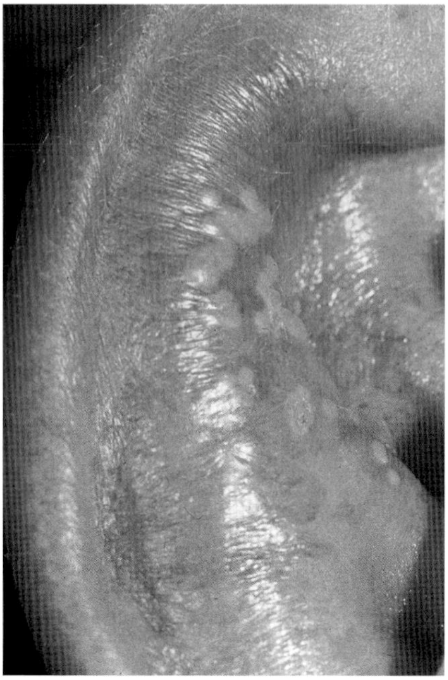

Fig. 1. Herpetiform eruptions of the auricle.

Non-neural satellite cells seem to play a major role as hosts for the virus [7]. Reactivation of the virus and clinical onset of the disease is typically characterized by an inflammatory infiltration – mostly lymphocytic or with other round cell elements – of the ganglion and/or its environment [7].

Facial palsy as a predominant clinical symptom may be related to a (necrotizing) inflammation of the ganglion. In some cases, however, it is caused by a neuritis or perineuritis of the nerve itself, sometimes without any involvement of the ganglion (fig. 2). Histological studies in temporal bones of decedents within the course of the disease revealed various perineural, perivascular or intraneural infiltration with different degrees of tissue destruction.

Secondary effects like swelling by edema and/or hemorrhage may cause compression of the nerve within its bony canal. In those areas blood circulation may severely be impaired, leading to additional damage.

More recent investigations show that viral infection and inflammation is not limited to the geniculate ganglion or the facial nerve, but may also be present in many other structures of the inner ear region. Thus VZV DNA was identified in or around the auditory nerve, the vestibular nerve, the modiolus of the cochlea, the organ of Corti, the vestibular ganglion, the maculae of the utricle and saccule, even the auricular muscles and the middle ear mucosa [7, 9].

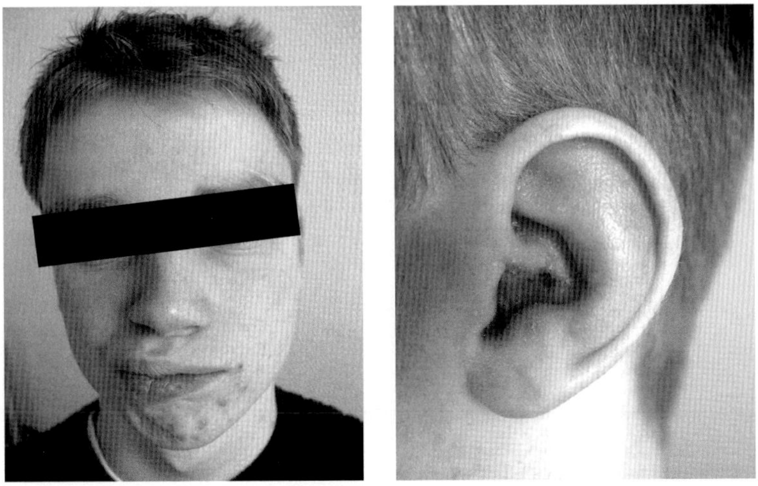

a b

Fig. 2. ***a*** Facial palsy. ***b*** Zoster oticus left ear.

Inflammation of the inner ear structures may explain otoneurological symptoms like sensorineural hearing loss or vertigo, which are frequently found in zoster oticus (fig. 3).

Possible routes of viral spread or 're-infection' lead from the facial nerve or the geniculate ganglion to the eighth cranial nerve via the anastomotic communition nerve branch or even from the infected middle ear mucosa to the inner ear through the round or oval window. However, determining the exact pathways seems problematic and not very relevant with regard to the fact, that many more neural structures (IX, X cranial nerve, brainstem, cerebellum, cerebrum) or fluid compartments (blood, liquor, saliva etc.) may be involved in the viral disease [7–11].

Clinical Symptoms

In general, zoster oticus is a disease of elderly people or patients with some kind of immune deficiency. It may however be observed in younger patients in stress conditions [12]. Even in children with acute peripheral facial paralysis, varicella zoster virus reactivation is considered an important factor [13]. The outbreak of the disease can sometimes even be related to surgical manipulations: there are publications of zoster oticus in consequence of dental or oropharyngeal surgery [14], after mandibular block [15], acoustic neuroma [16] – or even tympanoplastic middle ear surgery [17]. Very rarely the disease is found bilaterally in immunocompromised patients [18].

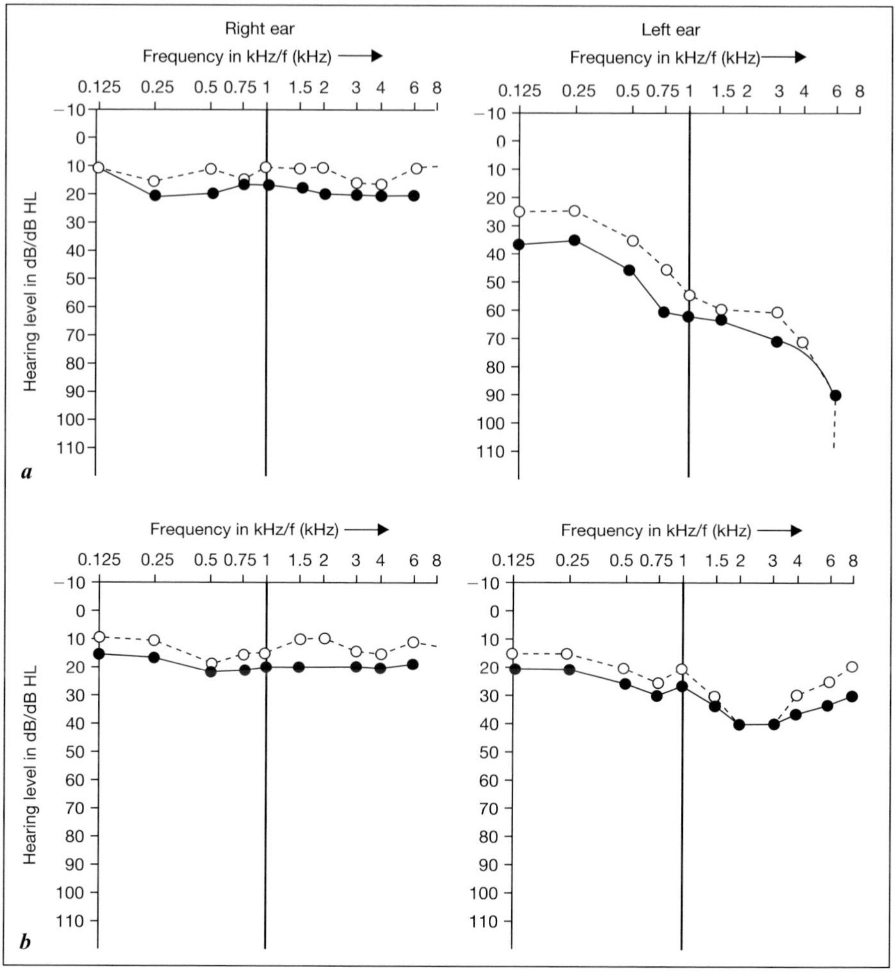

Fig. 3. a Audiogram of zoster-induced hearing loss (sensory-neural) of the left ear. ***b*** Almost complete recovery after 4 weeks (treatment with infusions plus acyclovir and cortisone).

The clinical picture may vary in severity from case to case. General characteristics are periauricular pain, herpetiform eruptions and neural dysfunctions, predominantly involving the VII and VIII cranial nerve [3, 19].

A prodromal stage with fatigue and feeling sick (approximately 7–14 days) is followed by a phase of erupting herpetiform lesions. Erythematous maculopapules around or on the auricle (most often in the concha or the superficial part of the outer ear canal) soon vesiculate and sometimes turn to ulcers. In rare cases, vesicles can (also) be observed in the buccal mucosa of the corresponding side or in the oropharynx. Body temperature may slightly be raised, usually there is pain – mostly moderate to severe – in the ear and its environment. After approximately 1 week, the lesions turn dry and crust. Regional lymphadenopathy is rather common. One of the leading symptoms is local pain (deep earache till pain around the auricle), which may persist even after the crust fell off ('postherpetic neuralgia').

Cutaneous zoster eruptions are in no way obligatory: 8–15% of the patients present without any rash or of vesicles ('zoster sine herpete' [20]).

Special attention must be turned to neural disorders: about 70% of the zoster oticus patients show some kind of involvement of the facial nerve, the incidence of inner ear affections (deafness, vertigo) seems to range within equal dimensions.

Signs of facial palsy mostly emerge later than the rash, typically proceeding rapidly from mild facial weakness to a complete unilateral peripheral paralysis.

Some patients complain of dysgeusia due to loss of chorda tympani-related functions. So unilaterally taste on the antero-lateral portion of the tongue may be diminished.

Without adequate treatment, the prognosis of facial palsy is very poor. Untreated, the chance of a restitutio ad integrum is as little as 10% in complete and about 70% incomplete paralysis [21]. With the nowadays generally accepted treatment the outcome is much more favorable.

Zoster-related hearing disorders may occur at different intervals in the course of the disease, but mostly follow the cutaneous symptoms. Often, there is an abrupt onset, resembling sudden hearing loss. The common finding is a sensorineural deafness, mostly in the high frequency range, sometimes accompanied by tinnitus. Profound inner ear hearing loss or even complete deafness is a very rare event. In some cases, so-called hyperacousis is found, meaning that the pain threshold for noise or tones is decreased, though (mostly) the hearing threshold is raised.

Vertigo may come along with the hearing disorders or appear at intervals, often with sensations of spinning or twirling motion.

Zoster oticus infection may cause neurologic complications such as cerebrospinal fluid changes, involvement of other cranial nerves (V, IX, X), aseptic meningitis, myelitis, encephalitis, acute ascending polyradiculitis, thrombotic cerebral vasculopathy etc. There are publications describing extensive involvement of more cranial nerves such as XI and XII [22], very rarely I, II, III and IV, cervical dermatomes C2–4 or even bilateral facial paralysis [18, 23]. In those cases, the spread of infection through meningeal inflammation should be considered.

Post-herpetic neuralgia can persist for long time after the disease, meaning a special therapeutic challenge.

Diagnosis

In cases with characteristic pathology (pain, herpetiform eruptions of the auricle, facial palsy, may be unilateral hearing loss, vertigo) the diagnosis may be fairly easy. However, there are many atypical courses, especially those without any cutaneous symptoms ('zoster sine herpete'), in which only serologic examinations can verify the diagnosis 'zoster'.

Serological assays confirming the diagnosis include verification of significant increase of Anti-VZV-antibodies of the IgG-, IgM-, respectively, IgA-fraction (Enzym-linked immunosorbent assay or immune fluorescence).

In some cases (and for scientific purpose), detection of viral DNA (VZV DNA) via polymerase chain reaction in cerebrospinal fluid, saliva tissue etc. can be helpful [24, 25].

In case of facial nerve involvement, the severity code is classified by the criteria of House and Brackmann [26]. The course of facial nerve palsy can be followed electrophysiologically (electromyography, electroneurography).

Hearing loss in the context of zoster oticus infection is documented and controlled by audiometry (pure tone audiometry, supra-threshold tests, impedance audiometry). In certain cases, objective audiometry via evoked potentials (e.g. BERA) can be helpful for identifying the site of the lesion (cochlear/retrocochlear) [27, 28]. In early stages registration of otoacoustic emissions can make sense [29].

In cases of zoster-linked dizziness or vertigo [30, 31], detection and registration of nystagmus is important. For clinical purpose, observation of eye movement, e.g. with special glasses ('Frenzel's goggles') in a darkened room may be sufficient. For more detailed examinations nystagmography (electronystagmography, or video-oculography) is inevitable (fig. 4). Caloric testing with water [32] for detecting unilateral vestibular disorders may be painful because of zoster papules, even the use of heated or cooled air may not be tolerated in single cases.

Additional hints concerning the site of the zoster-linked lesion may be given by magnetic resonance tomography of the temporal bone (fig. 5). Typically, gadolinium enhancement in the course of the facial nerve, the area of the geniculate ganglion, the labyrinth or the meatus acousticus internus may indicate inflammation within the temporal bone and help estimating the main focus of the disease [33–35]. Some authors advocate transcranial magnetic stimulation of the nerve, by which a more precise localization of the lesion within the Fallopian channel is possible. Moreover, additional electric stimulation via the same route may help to establish correct diagnosis and prognosis [36].

If there are further neurological signs or any apprehension of meningitis, lumbal puncture should be performed for examining CNS fluid.

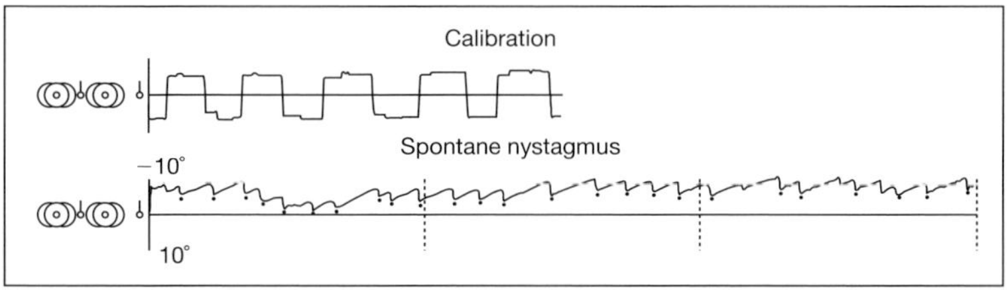

Fig. 4. Electronystamography presenting signs of spontaneous nystagmus in consequence of zoster oticus. The 'sawteeth' of the curve indicate spontaneous rhythmic eye movements, indicating unilateral peripheral vestibular dysfunction.

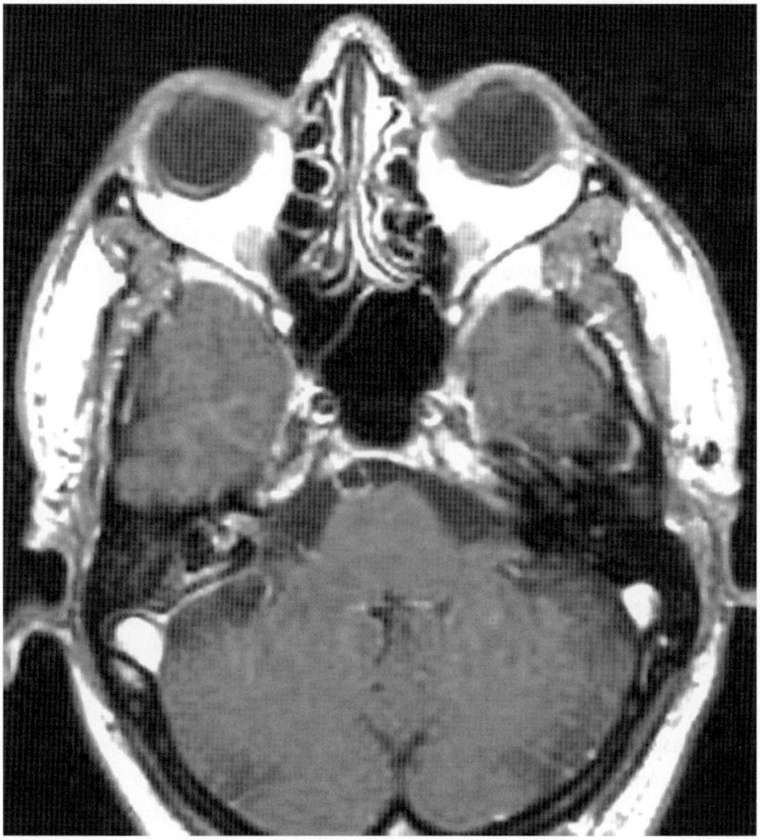

Fig. 5. MRI of the temporal bone showing zoster-related gadolinum enhancement of the intratemporal portion of the facial nerve.

Therapy

As soon as the diagnosis zoster oticus has been made, systemic antiviral chemotherapy should be started urgently. Substances like acyclovir, valacyclovir, famciclovir and brivudin are available. In most studies, high doses of acyclovir were administered intravenously (e.g. 5–10 mg, 3 times a day for 7–10 days), with good results [37, 38]. So far, in Germany oral application of antiviral agents, even brivudin as a very potent drug, has not been widely accepted for zoster oticus treatment, though in other localization of zoster it proved to be effective.

Once there are any signs of involvement of the VIIth or VIIIth cranial nerve (facial weakness, hearing loss, tinnitus, dizziness/vertigo), a simultaneous application of cortisone is indicated (e.g. 60 mg prednisolone daily for the first 10 days, subsequently 10 mg for another 10 days or starting with 200 mg prednisolone i.v., progressively decreasing till 10 mg p.o. within 2 weeks).

It seems crucial, not to delay the onset of the combined therapy (at least stay within the 72 h range), because the prognosis deteriorates markedly [39, 40].

With this therapy, prognosis of facial palsy has improved so much, that indications for surgical procedures for the nerve became very rare [41]. Some decades ago, nerve decompression for stopping 'self-strangulation' of the edematous nerve by removing the bony walls of its canal was a rather common operation [42]. Surgery ranged from simple mastoidectomy (descending part of the nerve) to transtemporal approaches, from which the entire course of the nerve from the internal auditory canal till the foramen stylomastoideum could be exposed [42, 43].

If the VIIIth cranial nerve is involved (hearing loss, tinnitus, dizziness/vertigo), a simultaneous infusion therapy (similar to that in sudden hearing loss) can be added. In Germany, the combination of plasma volume expanders (like hydroxyethyl starch) and vasoactive substances (e.g. pentoxiphylline) is widely used (e.g. 500 ml HAES plus pentoxiphylline per day for approximately 1 week).

In cases of vestibular imbalance with dizziness and vertigo sometimes causing vomitus, additional treatment with antivertiginous drugs: e.g. anticholinergics, antihistamines, Ca-channel blockers may be indicated.

Heavy pain is treated with analgesics (e.g. tramadol) in appropriate doses, best combined with a neuroactive agent (e.g. amitriptyline) [44].

Local treatment of the skin includes antisepsis and drying in the early stages (e.g. lotio alba, vioform zinc mixture), later cautious removal of crusts.

Prognosis

Prior to the era of early and simultaneous antiviral and antiphlogistic therapy, zoster-related affections of the VIIth and VIIIth cranial nerves had a poor prognosis. Nowadays, in case of early onset of medical treatment (as possible within the first 72 h), in most cases an almost or complete restitution can be expected. The main difficulty for the otologist is the correct interpretation of clinical symptoms, which may be misleading at least in the early phase of the disease. As indicative cutaneous symptoms may be missing ('zoster sine herpete'), the combination of local pain (earache) and hearing loss (sensorineural), vertigo or any kind of facial weakness should always be suspicious for zoster oticus.

References

1 Hunt JR: On herpetic inflammations of the geniculate ganglion. A new syndrome and its complications. J Nerv Ment Dis 1907;34:73–96.
2 Körner O: Über den Herpes zoster oticus. Munch Med Wochenschr 1904;1:6–10.
3 Walther LE, Prosowsky K, Walther A, Gudziol H: Herpes zoster oticus: symptom constellation and serological diagnosis. Laryngorhinootologie 2004;83:355–362.
4 Straus SE: Overview: the biology of varicella-zoster virus infection. Ann Neurol 1994;35(suppl): 4–8.
5 Gilden DH, Mahalingam R, Dueland AN, Cohrs R: Herpes zoster: pathogenesis and latency. Prog Med Virol 1992;39:19–75.
6 Gershon AA (ed): First international conference on the varicella-zoster virus. J Infect Dis 1992;166(suppl 1):1–68.
7 Wackym PA: Molecular temporal bone pathology: II. Ramsay Hunt syndrome (herpes zoster oticus). Laryngoscope 1997;107:1165–1175.
8 Meier JL, Straus SE: Comparative biology of latent varicella-zoster virus and herpes simplex virus infections. J Infect Dis 1992;166:13–23.
9 Murakami S, Nakashiro Y, Mizobuchi M, Hato N, Honda N, Gyo K: Varicella-zoster virus distribution in Ramsay Hunt syndrome revealed by polymerase chain reaction. Acta Otolaryngol 1998;118:145–149.
10 Gold E: Serologic and virus isolation studies of patients with varicella or herpes zoster infection. N Engl J Med 1966;274:181–185.
11 Bessho K, Ogino S: Detection of varicella-zoster virus DNA by polymerase chain reaction in patients with Hunt syndrome and clinical progression. J Otolaryngol Jpn 1994;97:1019–1027.
12 Meister W, Neiss A, Gross D, Doerr HW, Höbel W, Malin JP, von Essen J, Reimann BY, Witke C, Wutzler P: Demography, symptomatology and course of disease in ambulatory zoster patients. Intervirology 1998;41:272–277.
13 Furuta Y, Ohtani F, Aizawa H, Kukuda S, Kawabata H, Bergstrom T: Varicella-zoster virus reactivation is an important cause of acute peripheral facial paralysis in children. Pediatr Inf Dis J 2005;24:97–101.
14 Furuta Y, Ohtani F, Fukuta S, Inuymama Y, Nagashima K: Reactivation of varicella-zoster virus in delayed facial palsy after dental treatment abd oro-facial surgery. J Laryngol Otol 2000;62:42–45.
15 Maina S, Preece M: Herpes zoster oticus following mandibular block. J Laryngol Otol 2000;114: 212–213.
16 Magliulo G, D'Amico R, Celebrini A, Cuiuli G: Postoperative Ramsay-Hunt syndrome after acoustic neuroma resection. Viral reactivation. An Otorrinolaringol Ibero Am 2005;32:253–259.

17 Gyo K, Honda N: Delayed facial palsy after middle-ear surgery due to reactivation of varicella-zoster virus. J Laryngol Otol 1999;113:914–915.
18 Syal R, Tyagi I, Goyal A: Bilateral Ramsay Hunt syndrome in a diabetic patient. BMC Ear Nose Throat Disord 2004;4:3.
19 Robillard RB, Hilsinger RL, Adour KK: Ramsay Hunt facial paralysis: clinical analyses of 185 patients. Otolaryngol Head Neck Surg 1986;95:292–297.
20 Rudra T: Zoster sine herpete. Br J Clin Pract 1990;44:284.
21 Devriese PP, Moesker WH: The natural history of facial paralysis in herpes zoster. Clin Otolaryngol 1988;13:289–298.
22 De S, Pfleiderer AG: An extreme and unusual variant of Ramsay Hunt syndrome. J Laryngol Otol 1999;113:670–671.
23 Hamado T, Takano A, Miyao S, Teramoto J, Wanibe S: Bilateral facial palsy following trigeminal zoster with zoster oticus. Intern Med 2000;39:80–81.
24 Furuta Y, Ohtani F, Sawa H, Fukuda S, Inuyama Y: Quantification of varicella-zoster virus DNA in patients with Ramsay Hunt syndrome and zoster sine herpete. J Clin Microbiol 2001;39: 2856–2859.
25 Hiroshige K, Ikeda M, Hondo R: Detection of varicella virus DNA in tear fluid and saliva of patients with Ramsay Hunt syndrome. Otol Neurotol 2002;23:602–607.
26 House JW, Brackmann DE: Facial nerve grading system. Otolaryngol Head Neck Surg 1985;93: 146–147.
27 Abramovich S, Prasher DK: Electrocochleography and brain-stem potentials in Ramsay Hunt syndrome. Arch Otolaryngol Head Neck Surg 1986;112:925–928.
28 Wayman DM, Pham HN, Byl FM, Adour KK: Audiological manifestations of Ramsay Hunt syndrome. J Laryngol Otol 1990;104:8.
29 Kaberos A, Balatsouras DG, Korres SG, Kandiloros D, Economou C: Audiological assessment in Ramsay Hunt syndrome. Ann Otol Rhinol Laryngol 2002;111:68–76.
30 Proctor L, Perlmann H, Lindsay J, Matz G: Acute vestibular paralysis in herpes zoster oticus. Ann Otol Rhinol Laryngol 1979;88:303–310.
31 Kuhweide R, van de Steene V, Vlaminck S, Casselman JW: Ramsay Hunt syndrome: pathophysiology of cochleovestibular symptoms. J Laryngol Otol 2002;16:844–848.
32 Lu YC, Young YH: Vertigo from herpes zoster oticus: superior or inferior vestibular origin? Laryngoscope 2003;113:307–311.
33 Casselmann JW, Kuhweide R, Ampe W, Meeus L, Steyaert C: Pathology of the membraneous labyrinth: a comparison of T1- and T2-weighted and gadolinium-enhanced spin-echo and 3DFT-CISS imaging. Am J Neuroradiol 1993;14:59–69.
34 Korzec K, Sobol SM, Kubal W, Mester SJ, Winzberg G, May M: Gadolinium-enhanced magnetic resonance imaging of the facial nerve in zoster oticus and Bell's plasy: clinical implications. Am J Otol 1991;12:163–168.
35 Brandle P, Saretti-Schefer S, Bohm A, Wichmann W, Fisch U: Correlation of MRI, clinical and electroneuronographic findings in acute facial palsy. Am J Otol 1996;17:154–161.
36 Nowak DA, Linder S, Topka H: Diagnostic relevance of transcranial magnetic and electric stimulation of the facial nerve in the management of facial palsy. Clin Neurophysiol 2005;116: 2051–2057.
37 Dickens JRE, Smith JT, Graham SS: Herpes zoster oticus: treatment with intravenous acyclovir. Laryngoscope 1988;98:776–779.
38 Kinishi M, Amatsu M, Mohri M, Saito M, Hasegawa T, Hasegawa S: Acyclovir improves recovery rate of facial nerve palsy in Ramsay Hunt syndrome. Auris Nasus Larynx 2001;28:223–226.
39 Uri N, Greenberg E, Meyer W, Kitzes-Cohen R: Herpes zoster oticus: treatment with acyclovir. Ann Otol Rhinol Laryngol 1992;101:161–162.
40 Murakami S, Hato N, Horiuchi J, Honda N, Gyo K, Yanagihara N: Treatment of Ramsay Hunt syndrome with acyclovir-prednisone: significance of early diagnosis and treatment. Ann Neurol 1997;41:353–357.
41 Ko JY, Sheen TS, Hsu MM: Herpes zoster oticus treated with acyclovir and prednisolone: clinical manifestations and analysis of prognostic factors. Clin Otolaryngol 2000;25:139–142.

42 Fisch U, Esslen E: Total intratemporal exposition on the facial nerve. Arch Otolaryng 1972;95: 335–341.
43 Fisch U: Surgery for bell's palsy and herpes zoster oticus. Proceedings of the Symposium of Neurological Surgery of the Ear. Sarasota, 1979.
44 Malin JP: Zoster und Nervensystem. Dtsch Med Wochenschr 1996;121:635–638.

Prof. Dr. H. W. Pau
Klinik und Poliklinik für Hals-Nasen-Ohrenheilkunde
Kopf- und Hals-Chirurgie, 'Otto Körner' der Universität Rostock
Doberaner Strasse 137–139
DE–18057 Rostock (Germany)
Tel. +49 381 494 8300, Fax +49 381 494 8302, E-Mail hans-wilhelm.pau@med.uni-rostock.de

Gross G, Doerr HW (eds): Herpes Zoster.
Monogr Virol. Basel, Karger, 2006, vol 26, pp 58–68

Neuroanatomy of Pain and Neuropathology of Herpes Zoster and Postherpetic Neuralgia

Andreas Wree[a], *Oliver Schmitt*[a], *Kamen G. Usunoff*[a–c]

[a]Institute of Anatomy, Medical Faculty, University of Rostock, Rostock, Germany; [b]Department of Anatomy and Histology, Preclinical University Center, Medical University of Sofia, [c]Institute of Physiology, Bulgarian Academy of Sciences, Sofia, Bulgaria

Pain is an important biological signal for danger and necessary for the integrity of the organism. There are two different kinds of pain. One kind is typical for threat of tissue damage (pinprick sensation). It is rapidly conducted to consciousness and well localized. The other kind of pain occurs when tissue damage has already taken place. It is slowly conducted and poorly localized. But sustained or chronic pain can result in secondary symptoms. A typical example is postherpetic neuralgia (PHN), characterized by chronic and severe intractable pain [1–3].

Functional Neuroanatomy of the Pain System

Primary Afferent Neuron

The primary afferent (PA) neuron is a pseudounipolar cell, localized in spinal (dorsal root) ganglia (SG), and in sensory ganglia of the Vth, VIIth, IXth and Xth cranial nerves. Their perikarya are round to elliptical and emit a single process that bifurcates in a peripheral and central process. The central one enters the central nervous system (CNS), and the peripheral one runs to its zone of sensory innervation. There are two basic types of pseudounipolar perikarya: large, light A-cells and small, dark B-cells. The largest A-cells are the proprioreceptor neurons, and the small B-cells are the typical nociceptor neurons [4, 5]. Their transmitter is glutamate. In addition, the B-cells contain various neuropeptides, especially substance P (SP) [6, 7]. The peripheral processes of the

B-cells are thin fibers of two types, both terminating as free nerve endings [8]: Aδ-fibers are thinly myelinated (diameter of 1–3 μm, conduction velocity of 5–30 m/s), and C-fibers are unmyelinated (diameter of 1 μm, conduction velocity of only 0.5–2 m/s). Stimulation of Aδ-nociceptors leads to pricking, that of C-nociceptors to burning or dull pain [9]. The free nerve endings are found throughout the body, mainly in the adventitia of small blood vessels, outer and inner epithelia, connective tissue capsules and the periosteum. They are most densely in the cornea, tympanic membrane, dental pulp, skin and mucosa of the head, skin of the fingers, parietal pleura and peritoneum. Except for avascular structures, the nociceptors are adjacent to capillaries and mast cells. This triad is a functional nociceptive response unit [7]. The firing of nociceptors at the site of tissue injury causes release of vesicles containing peptides that act in sensitizing the nociceptor and increase its firing rate [7]. Cellular damage and inflammation increase concentrations of other chemical mediators (histamine, bradykinin, prostaglandins) also, augmenting the transmission of nociceptive impulses along sensory afferent fibers [7]. In addition, pronociceptive roles have been proposed for 'exotic' species, including protons, ATP, cytokines, neurotrophins and NO [9]. There are nociceptors that normally are inactive and rather unresponsive. Inflammation leads to sensitization of these fibers, becoming sensitive to peripheral stimulation [10].

Termination of PA Fibers in the Spinal Cord and Spinal Trigeminal Nucleus

As central processes of SG neurons approach the dorsal root entry zone, the nociceptive axons enter lateral portions of the dorsal horn (DH). They terminate primarily in laminae I and II (substantia gelatinosa) [11], Aδ-fibers in laminae I and V, C-fibers in lamina II. The mechanoreceptive Aβ-axons reach laminae III–VI [4, 12]. The large lamina I neurons are the source of about one-half of the spinothalamic tract (STT). Lamina II contains small cells with rich dendritic trees. In lamina II neurons the inhibitory transmitters γ-aminobutyric acid and glycine coexist. Lamina II functions as a controlling system modulating synaptic transmission from PA neurons to secondary sensory systems. After complex local processing in the DH [4], nociceptive signals are conveyed to higher centers through projection neurons whose axons form several ascending fiber systems. The central processes of trigeminal ganglion neurons enter the brainstem via the trigeminal root. Some fibers bifurcate to give a rostral branch to the principal trigeminal nucleus and a caudal one that joins the spinal trigeminal tract reaching the ipsilateral spinal trigeminal nucleus (STN) [5]. The PA fibers terminate somatotopically: the ophthalmic fibers are located ventrically, the maxillary in the middle and the mandibular fibers are found dorsally. The PA axons emit collaterals to all three subnuclei of the STN (oralis, interpolaris and caudalis), and nociceptive Aδ- and C-fibers terminate in subnucleus caudalis. The laminar structure of subnucleus caudalis of the STN is very similar to the spinal DH.

Ascending Pathways of the Spinal Cord and of the Spinal Trigeminal Nucleus Spinothalamic Tract. The STT mediates the sensations of pain, cold, warmth, and touch [13]. The cells of origin are located in laminae I, IV–VI [12, 14], their transmitter is glutamate. Most cells project to the contralateral thalamus. Clinical observations, however, indicate that ipsilaterally projecting STT neurons also exist [13]. The STT axons cross the midline in the commissura alba anterior and ascend in the anterolateral quadrant of the white matter somatotopically arranged. The STT axons terminate in the nucleus ventralis posterior lateralis (VPLc and VPLo), the nucleus ventralis posterior inferior (VPI), the medial part of the posterior nuclear complex (Pom) as well as in intralaminar and medial thalamic nuclei. STT neurons projecting to the lateral thalamus (VPL) have receptive fields on a restricted area. Therefore, they are suited for signaling the sensory-discriminative aspects of pain (fig. 1).

Trigeminothalamic Tract. The entire trigeminal nuclear complex projects bilaterally but strongly crossed to the thalamus. The projection to VPM and the posterior thalamic nucleus (Po) arise in the principal trigeminal nucleus and in subnucleus interpolaris, while nucleus submedius and the intralaminar nuclei are heavily innervated by the nociceptive subnucleus caudalis. The lamina I neurons send strong projections to nucleus submedius, VPM, and Po. The deeper laminae innervate moderately VPM and Po, but project heavily to the ventral diencephalon.

Further Ascending Pathways. Several other pathways accompany the STT in the ventrolateral quadrant of the spinal cord. These are the spinomesencephalic tract (SMT), the spinoparabrachial tract, the spinoreticular tracts, and several spinolimbic tracts [4, 15]. The SMT actually includes projection systems that terminate in different mesencephalic areas. The neurons of origin are distributed similar to the STT neurons. The bilateral SMT projections terminate in the periaqueductal gray (PAG) and other nuclei [4]. Although the SMT is involved in nociception, it seems to contribute to the motivational affective aspects of pain, and triggers activity in the descending control systems. Spino(trigemino) parabrachial tract is a major nociceptive projection, rivalling with the STT. The parabrachial nucleus is densely innervated by ascending spinal cord and STN axons. The cells of origin are located in lamina I, e.g. receiving a nociceptive input from SP-releasing PA neurons. The spinoparabrachial tract is bilateral. The parabrachial nucleus projects heavily to the amygdala and the hypothalamus. The

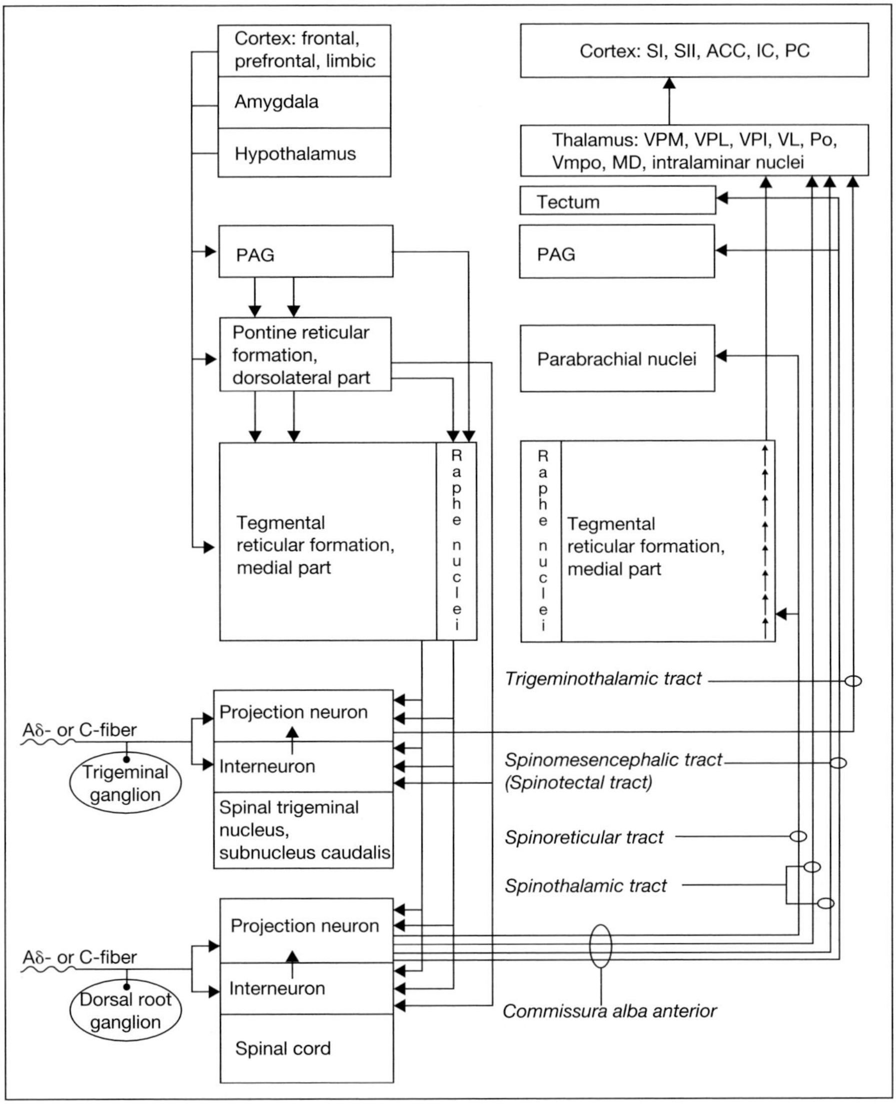

Fig. 1. Schematic diagram of important ascending and descending pathways of the nociceptive system. ACC = Anterior cingulate; IC = insular; MD = mediodorsal; PAG = periaqueductal gray; PC = prefrontal; Po = posterior thalamic nucleus; SI = primary somatosensory; SII = secondary somatosensory; VL = nucleus ventralis lateralis; Vmpo = nucleus ventralis medialis, posterior division; VPL = nucleus ventralis posterior lateralis; VPI = nucleus ventralis posterior inferior.

spino–parabrachio–amygdalar/hypothalamic nocispecific multineuronal chain is probably concerned with the intensity of pain. The spino–reticulo–thalamic pathways may play an important role in distributing pain signals to the forebrain. In addition, there are direct spino(trigemino)hypothalamic and spino(trigemino) limbic tracts. The spinohypothalamic tract is an unexpectedly massive projection, terminating in most of the hypothalamic divisions. Thus, the spinohypothalamic tract provides nociceptive input to many areas that are involved in the multifaceted responses to noxious stimuli.

Cerebral Cortices Involved in Pain Perception

There is a multiregional organization of cortical pain processing [16] and areas involved are the primary (SI) and the secondary (SII) somatosensory, the insular (IC), the anterior cingulate (ACC) and the prefrontal (PC) cortices. These areas differ functionally as seen in electrophysiological and functional imaging studies: the sensory-discriminative aspect of pain (localization, intensity, duration, quality) is presented in SI (and SII), receiving input from lateral thalamic nuclei, the motivational-affective aspect (subjective suffering, unpleasantness, aversive emotions) and the cognitive-evaluative aspects of pain are presented in IC, ACC and PC, receiving input from medial thalamic nuclei.

SI is Located in the Postcentral Gyrus. Two classes of neurons are activated in SI: neurons with a wide dynamic range react already to not painful stimuli, showing highest activity to painful stimuli, having large receptive fields and coding pain intensity; specific nociceptive neurons react to painful stimuli only, having small receptive fields, being somatotopically located and enabling localization, intensity and temporal attributes of painful stimuli. The SI neurons get their afferences from the lateral thalamic nuclei (VPL, VPM, VPI) and also heavily project back to them. Lesions of the respective thalamic nuclei, the thalamo-cortical connections or of SI result in a decrease in temperature and pain perception. But there is no complete analgesia, pain is still interpreted as uncomfortable [17].

SII is Located just Lateral and Slightly Anterior to the Lateral end of the Central Fissure. SII neurons seemingly are not involved in discrimination of location and/or intensity of painful stimuli, but have an important role in recognition, learning and memory of painful events [18]. There is significant pain-related activation of SII [19]. The SII neurons get their bilateral afferences from the lateral thalamic nuclei partly different from those projecting to SI, namely from VPI and the dorsal part of Po. Thus, SII may play a role in relaying nociceptive information to IC and temporal lobe limbic structures providing fast access to pain related learning and memory.

Functional Imaging Studies Show an Increased Blood Flow of the IC during Painful Stimuli, Either Contralaterally or Bilaterally [19]. Patients with lesions of the IC had an elevated pain tolerance and loss of or inadequate emotional reactions to painful stimuli although recognizing pain (asymbolia for pain) [20]. The IC gets thalamic afferents from the VMpo, mediodorsal (MD) and intralaminar nuclei and from SII, and projects to the amygdala, speaking in favor of the importance of IC in the motivational-affective aspect of pain and in autonomic reactions to noxious stimuli.

The ACC is Involved in Cognition and Emotion. The subarea involved in the motivational-affective aspect of pain is located in the rostral part of Brodmann's area 24 and the adjoining area 32. Patients with ACC lesions lost the emotional reactions to painful stimuli although pain could be correctly localized. In ACC pain receptive neurons were found with large, often bilateral receptive fields. The ACC gets thalamic afferents from VMpo, MD and intralaminar nuclei, from IC and PC, and projects to the amygdala, MD, PAG, motor nuclei of the brainstem and IC. The ACC is involved in motivational-affective aspects of pain and in conditioned fear reaction, having a pivotal role in interrelating attentional functions with that of establishing emotional valence and response properties [21].

There are Still Doubts with Respect to the Function of the PC in Pain Perception. The PC functions as 'supervisory attention system' are correlated with the cognitive-evaluative aspect of pain. The PC gets thalamic afferents from VMpo, MD and intralaminar nuclei, and projects to MD and ACC. Functional imaging studies show activation of parts PC (probably Brodmann's areas 9 and 10) during painful stimuli, mostly in the right hemisphere. Patients with unilateral lesions of the PC show changes in both the sensory-discriminative and the motivational-affective aspects of pain.

Descending Modulatory Pathways

The communication of Reynolds [22] being able to perform abdominal surgery in rats by stimulating the midbrain PAG, initiated investigations on the 'descending analgesia systems'. The PAG neurons project to the serotoninergic raphe nuclei and to the noradrenergic nuclei in the dorsolateral pons. Both neuronal groups project heavily to the spinal cord and to the STN. The projections are predominantly crossed, innervating laminae I, II and V. The neurochemistry of the transmitters and receptors in the multineuronal antinociceptive pathway arising in the PAG is complex. Along serotonin and noradrenaline, also endogenous opiates and glutamate, γ-amino butyric acid and glycine are involved.

Herpes Zoster and Postherpetic Neuralgia

Neuropathology

CNS complications occur in less than 1% of chickenpox cases [23]. The children have mild meningitic symptoms. The most common abnormality is the cerebellar ataxia; very rarely transverse myelitis. After varicella resolves, VZV becomes latent in sensory ganglia, persisting throughout lifetime [23]. VZV is localized predominantly in the pseudounipolar PA neurons and/or in the perineuronal satellite cells. During latency, VZV is not infectious and does not transcribe most of its genetic material, thereby escaping from detection and clearance by the host immune system. The likelihood of viral reactivation to herpes zoster (HZ) increases with each advancing decade of age. Immunocomprised patients are at an especially high risk [23]. With reactivation, the virus spreads transaxonally to the skin, causing a rash with a dermatomal distribution, and severe radicular pain. Thoracic HZ is the most common one, affecting one to two, rarely more dermatomes, followed by the ophthalmic division of the trigeminal nerve [23]. The involvement of the facial nerve results in HZ oticus: geniculate neuralgia described by Ramsay Hunt [24]. Similar combination of painful dermatomal rash with myotomal motor weakness might occur also in spinal nerve HZ. Frequently, a prodrome of dermatomal pain starts before the appearance of the rash. Dermatomal pain without a rash ('zoster sine herpete') occurs rarely. HZ is mostly monophasic in immunocompetent patients. In contrast, in immunocompromized patients (especially in AIDS) HZ is recurrent, protracted, and often accompanied with severe neurological complications [25].

The basic neuropathological substrate for HZ is ganglionic hemorrhage, necrosis and inflammation [23]. The histopathologic features include mononuclear and lymphocytic infiltration, neuronal degeneration, neuronal phagocytosis by satellite cells, empty neuronal cell beds and fibrous scarring of the ganglia. Vasculitis in the adjacent nerve results in damage of axons and destruction of myelin sheaths [26]. Rarely, VZV spreads in the CNS [23] causing myelitis. In HZ ophthalmicus, the virus might spread via trigeminal afferent fibers to the large blood vessels at the base of the brain, with resultant vessel thrombosis, vessel wall inflammation, and large, ipsilateral brain infarctions. Mostly, HZ resolves without sequelae. However, many elderly patients have prolonged, debilitating pain, known as PHN [25]. Besides PHN, also other peripheral nerve injuries (cf. traumatic injury, diabetes, malignancy) might result in neuropathic pain (NP). NP conditions are characterized by: spontaneous, continuous pain, usually of a burning character, paroxysmal (lancinating); evoked pain to mechanical or thermal stimuli, such as allodynia and hyperalgesia. Allodynia can be produced in two ways: by the action of low threshold myelinated Aβ-fibers on an altered CNS, and by a reduction in the threshold of nociceptive fibers in the periphery. The fact that pain is often located in hypoesthetic or anesthetic areas appears paradoxical and implies that NP not only depends on the messages from nociceptors, but also on other mechanisms as well, in contrast to 'nociceptive pain' [27].

The pathology of PHN is just beginning to be understood [23]. Smith [28] found cystic distortion of thoracic SG removed 2.5 months after the onset of HZ, and persistent chronic inflammatory cells. In a patient with removed SG 2 years after the onset of PHN he found 'ghost cells', hypothesizing that the altered structure of surviving cells might contribute to the intractable pain. Furthermore, DH atrophy and cell, axon and myelin loss were encountered in PHN patients. Interestingly, a greater loss of small cutaneous nerve endings in skin biopsies obtained from patients with HZ who developed PHN was demonstrated [29].

Axons of uninjured PA neurons terminating in the DH can collaterally sprout following chronic NP [30]. Peripheral nerve injury results in a rearrangement of the highly ordered laminar termination of PA fibers within appropriate regions of the DH. Peripheral axotomy causes long-lasting sprouting of A-fibers into lamina II, an area they do not normally terminate. This A-fiber sprouting appears to be a result of two phenomena: firstly, the presence of vacant synaptic sites within the superficial laminae following the degeneration of C-axons, secondly the induction of a regenerative capacity in the injured neurons. The result of A-fiber sprouting is that lamina II begins to receive information about non-noxious stimuli, an information being misinterpreted by the CNS as noxious (mechanical allodynia). Thus, peripheral and central mechanisms contribute to PHN pain. Some PHN patients show abnormal sensitization of unmyelinated nociceptors (irritable nociceptors), having minimal sensory loss. Others have pain associated with small fiber deafferentiation: pain and temperature sensations are profoundly impaired but mechanical stimuli can produce severe pain (allodynia). In those, allodynia may be due to the formation of new connections between non-nociceptive thick (Aβ) PA fibers and central pain transmission neurons. The third class complains from severe spontaneous pain without hyperalgesia or allodynia, having lost both large and small diameter fibers, and pain is due to increased spontaneous activity in deafferented central neurons and/or reorganization of central connections. The central sensitization is an activity-dependent functional plasticity that results from activation of intracellular kinase cascades leading to increasing synaptic efficacy [30, 31]. One factor underlying the insensitivity of NP to opioid analgetics could be due to a marked reduction in the number of μ-opioid receptors both in the axotomized primary sensory and in the lamina II neurons. In consequence, due to the different neuropathological mechanisms obviously underlying PHN, appropriate treatment approaches in individual patients vary [32, 33].

Changes in Cortical Networks due to Chronic Pain

Chronic pain or NP can result from damage at different levels of pain processing: peripheral nerve, SG, dorsal root, CNS. Chronic syndromes show positive symptoms like pain, dysesthesia and paresthesia, often in combination with negative ones like sensory deficits. Peripheral neuropathic pain is a spontaneous stimulus-independent or a hypersensitivity pain caused by a stimulus following damage of sensory neurons (stimulus-evoked pain). Inflammation in the DG can sensitize neurons to respond to normal innocuous thermal or mechanical stimuli, and loss of DG perikarya can induce changes in surrounding surviving neurons. Thus, loss of sensory dendrites in the epidermis of patients suffering from PHN was positively correlated with both sensory deficits and with pain [29]. Changes caused by alterations of peripheral input, followed by changed spinal processing can be forwarded to the cortex via thalamic nuclei [34]. Neurons in the somatosensory thalamus of NP patients showed various electrophysiological abnormalities: responses to stimuli of body regions not normally driving those cells, high spontaneous firing rates, abnormal bursting activities. Thus, besides peripheral and spinal changes there is massive thalamic and cortical plasticity contributing to the development of pathological pain.

References

1 Dworkin RH, Johnson RW: A belt of roses from hell: pain in herpes zoster and postherpetic neuralgia; in Block AR, Kremer EF, Fernandez E (eds): Handbook of Pain Syndromes: Biopsychosocial Perspectives. New Jersey, Erlbaum, Hillsdale, 1999, pp 371–402.

2 Dworkin RH, Portenoy RK: Pain and its persistence in herpes zoster. Pain 1996;67:241–251.

3 Usunoff KG, Popratiloff A, Schmitt O, Wree A: Functional neuroanatomy of pain. Adv Anat Embryol Cell Biol 2006;184:1–119.

4 Willis WD, Coggeshall RE: Sensory Mechanisms of the Spinal Cord, ed 2. New York, Plenum Press, 1991.

5 Usunoff KG, Marani E, Schoen JH: The trigeminal system in man. Adv Anat Embryol Cell Biol 1997;136:I–X, 1–126.

6 Lazarov NE: Comparative analysis of the chemical neuroanatomy of the mammalian trigeminal ganglion and mesencephalic trigeminal nucleus. Prog Neurobiol 2002;66:19–60.

7 McHugh JM, McHugh WB: Pain: neuroanatomy, chemical mediators, and clinical implications. AACN Clin Issues 2000;11:168–178.
8 Heppelmann B, Messlinger K, Neiss WF, Schmidt RF: Fine sensory innervation of the knee joint capsule by group III and IV nerve fibers in the cat. J Comp Neurol 1995;351:415–428.
9 Millan MJ: The induction of pain: an integrative review. Prog Neurobiol 1999;57:1–164.
10 Schaible HG, Schmidt RF: Time course of mechanosensitivity changes in articular afferents during a developing experimental arthritis. J Neurophysiol 1988;60:2180–2195.
11 Rexed B: Some aspects of the cytoarchitectonics and synaptology of the spinal cord; in Eccles JC, Schade JP (eds): Organization of the Spinal Cord. Prog Brain Res, vol 2. Amsterdam, Elsevier, 1964, pp 58–92.
12 Craig AD: Pain, temperature, and the sense of the body; in Franzen O, Johansson R, Terenius L (eds): Somesthesis and the Neurobiology of the Somatosensory Cortex. Basel, Birkhäuser, 1996, pp 27–39.
13 Nathan PW, Smith M, Deacon P: The crossing of the spinothalamic tract. Brain 2001;124: 793–803.
14 Andrew D, Craig AD: Spinothalamic lamina I neurons selectively sensitive to histamine: a central neural pathway for itch. Nat Neurosci 2001;4:72–77.
15 Willis WD, Westlund KN: Neuroanatomy of the pain system and of the pathways that modulate pain. J Clin Neurophysiol 1997;14:2–31.
16 Peyron R, Laurent B, Garcia-Larrea L: Functional imaging of brain responses to pain: a review and meta-analysis. Neurophysiol Clin 2000;30:263–288.
17 Ploner M, Freund HJ, Schnitzler A: Pain affect without pain sensation in a patient with postcentral lesion. Pain 1999;81:211–214.
18 Schnitzler A, Ploner M: Neurophysiology and functional neuroanatomy of pain perception. J Clin Neurophysiol 2000;17:592–603.
19 Treede RD, Apkarian AV, Bromm B, Greenspan JD, Lenz FA: Cortical representation of pain: functional characterization of nociceptive areas near the lateral sulcus. Pain 2000;87: 113–119.
20 Greenspan JD, Lee RR, Lenz FA: Pain sensitivity alterations as a function of lesion location in the parasylvian cortex. Pain 1999;81:273–282.
21 Price DD: Psychological and neural mechanisms of the affective dimension of pain. Science 2000;288:1769–1772.
22 Reynolds DV: Surgery in the rat during electrical analgesia induced by focal brain stimulation. Science 1969;164:444–445.
23 Kleinschmidt-DeMasters BK, Gilden DH: Varicella-zoster virus infections of the nervous system. Clinical and pathologic correlates. Arch Pathol Lab Med 2001;125:770–780.
24 Hunt JR: On herpetic inflammations of the geniculate ganglion: a new syndrome and its complications. J Nerv Ment Dis 1907;34:73–96.
25 Gilden DH, Kleinschmidt-DeMasters BK, LaGuardia JJ, Mahalingham R, Cohrs RJ: Neurologic complications of the reactivation of varicella-zoster virus. N Engl J Med 2000;342:635–645.
26 Fabian WA, Wood B, Crowley P, Kakulas BA: Herpes zoster brachial plexus neuritis. Clin Neuropathol 1997;16:61–64.
27 Attal N, Bouhassira D: Mechanisms of pain in peripheral neuropathy. Acta Neurol Scand 1999;173(suppl):12–24.
28 Smith FP: Pathological studies of spinal nerve ganglia in relation to intractable intercostal pain. Surg Neurol 1978;10:50–53.
29 Oaklander AL, Romans K, Horasek S, Stocks H, Hauer P, Meyer RA: Unilateral postherpetic neuralgia is associated with bilateral neuron damage. Ann Neurol 1998;44:789–795.
30 Ji RR, Woolf CJ: Neuronal plasticity and signal transduction in nociceptive neurons: implications for the initiation and maintenance of pathological pain. Neurobiol Dis 2001;8:1–10.
31 Hunt SP, Mantyh PW: The molecular dynamics of pain control. Nat Rev Neurosci 2001;2:83–91.
32 Baron R: Post-herpetic neuralgia case study: optimizing pain control. Eur J Neurol 2004;11(suppl 1): 3–11.

33 Wasner G, Kleinert A, Binder A, Schattschneider J, Baron R: Postherpetic neuralgia: topical lidocaine is effective in nociceptor-deprived skin. J Neurol 2005;252:677–686.
34 Hsieh JC, Belfrage M, Stone-Elander S, Hansson P, Ingvar M: Central representation of chronic ongoing neuropathic pain studied by positron emission topography. Pain 1995;63:225–236.

Andreas Wree, MD
Institute of Anatomy, University of Rostock
PO Box 100888
DE–18055 Rostock (Germany)
Tel. +49 381 494 8401, Fax +49 381 494 8402, E-Mail andreas.wree@med.uni-rostock.de

Gross G, Doerr HW (eds): Herpes Zoster.
Monogr Virol. Basel, Karger, 2006, vol 26, pp 69–80

Postherpetic Neuralgia and Other Neurologic Complications

Ralf Baron

Neurological Clinic, Pain Research and Therapy, University Kiel, Kiel, Germany

Definition

The acute herpes zoster radiculoneuritis affects mainly elderly patients with an incidence of 125/100,000 per year. The clinical onset of acute herpes zoster infection is heralded by pain in the affected segment (preherpetic neuralgia). The characteristic vesicopapular rash usually appears a few days after the onset of pain and takes 3–4 weeks to heal. In most patients, the rash and pain disappear completely within a period of 1–2 months. These patients develop neither local neuropathy nor other cutaneous sensory changes. In other patients, the acute neurocutaneous symptoms may be followed by irreversible skin damage and sensory abnormalities and, in a significant number of patients, there is persistent pain or the initial pain subsides and a second pain, often of different character, begins. This condition is called postherpetic neuralgia (PHN). In the overall population on average 12–20% suffer from pain at the time of skin healing and 2–5% at 1 year after zoster. The incidence of PHN dramatically depends on the age of the patients. In the age population of 60–70 years 50–75% of the patients suffer from long-lasting PHN. The average age of the PHN patients is 70 years, there are no gender differences [1].

Clinical Characteristics

Basically, each dermatome can be affected by the zoster infection. However, there is a peak in thoracic dermatomes (54%), in particular TH5 (15%), and in the face (20%), in particular V1 (13%). In about 50% of the patients two or more segments are affected, there are no differences between the right and the left side [1].

Dermatological and Neurological Complications of Acute Herpes Zoster

Acute zoster in the first trigeminal distribution (ophthalmic zoster) very often leads to severe ocular complications (fig. 1). The cornea is very densely innervated by trigeminal afferent fibers. Therefore, in 25–70% the cutaneous zoster is associated with keratitis, iritis or chorioiritis which may lead to blindness.

If the facial nerve and the geniculate ganglion is affected by the acute zoster infection (otic zoster), the rash will appear in the outer ear tube and a peripheral facial palsy occurs that often has a poor long-term outcome.

If the acute zoster infection affects the ventral root or the ventral horn of the spinal cord, in addition to sensory symptoms a severe muscle weakness may develop. Does the zoster infection involve the brachial or lumbosacral plexus, in 1–5% of the cases proximal pareses of the skeletal muscles will occur (fig. 2). If the virus is transported centrally to the nervous system polyradiculitis, myelitis and encephalitis may occur [2].

In immuno-compromised patients generalization of the rash to the entire body (zoster generalisatus) is a common complications of acute zoster.

Pain Chronification – Postherpetic Neuralgia

Although histopathological studies have demonstrated peripheral and central degenerative changes in many PHN patients, i.e. ganglion cell loss, fibrosis and atrophy of the dorsal horn, dorsal root ganglion, dorsal root and peripheral nerve [3–7], the most incapacitating symptom of PHN is the nearly intractable pain and not just simple sensory loss. Most patients with PHN are able to distinguish three distinct components of their discomfort: (1) a constant, spontaneous usually deep burning pain, (2) a brief recurrent shooting tic-like pain and (3) a sharp radiating pain of burning character evoked by very light touching of the skin, which is called dynamic mechanical allodynia (evoked-pain). The patient may undergo extraordinary efforts to protect the diseased area from innocuous mechanical stimuli. Yet, firm compression of the skin mostly does not exacerbate the pain and may even provide relief. Clinical investigations show that negative (i.e. sensory deficits) and positive (i.e. mechanical allodynia) sensory signs may coexist within the affected dermatome. Extreme allodynia to light touch is often restricted to reproducable areas surrounding scarred skin or lying at the border of affected and unaffected dermatomes. In some patients the allodynic area expands far into formerly unaffected adjacent dermatomes already indicating that some central nervous system mechanisms must be involved in the generation of allodynia. Besides mechanically evoked pain types up to 30% of patients suffer from heat evoked pain (heat hyperalgesia), whereas pain induced by cold stimuli is a very rare phenomenon (<10%).

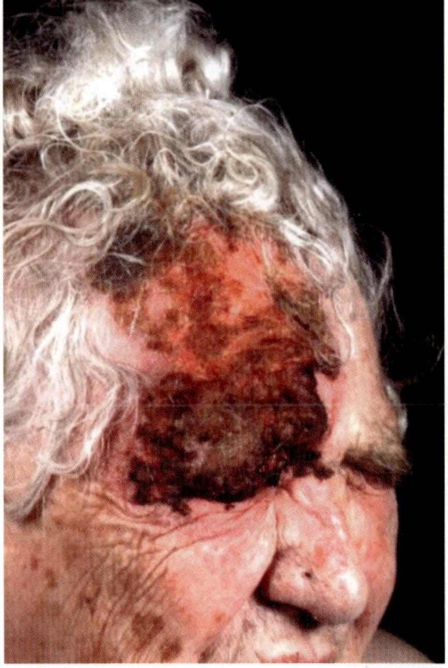

Fig. 1. Severe acute herpes zoster within the right first trigeminal distribution. The patient suffered from severe pain and visual loss. Involvement of the cornea which is densely innervated by trigeminal afferent fibers.

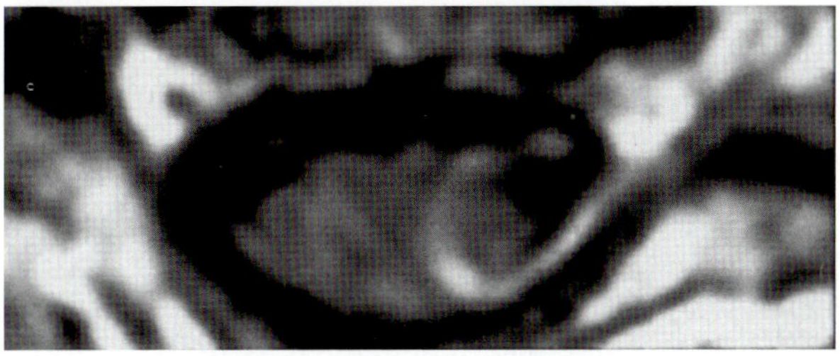

Fig. 2. Magnetic resonance image of a patient with acute herpes zoster. Spinal cord in the cervical affected region. Note the contrast enhancement in the dorsal root, dorsal horn of the spinal cord and lesser also in the ventral root. The patient was severely affected with pain and motor weakness at the left arm (T1 image with contrast medium). Photo courtesy of Prof. Dr. T. Tölle, Munich.

Pathophysiological Mechanisms of Pain Generation

Considerable advances in the understanding of chronic pain mechanisms have been achieved in the last years mainly using animal models with a variety of different nerve lesions and the more recent models of VZV infection [8, 9]. In the latter, the striking PHN symptoms of hyperalgesia and allodynia can be reproduced for a period of at least 3 months in vivo by injection of virus into the rat footpad [10].

These animal models are currently the major source of our knowledge on pathophysiological mechanisms in neuropathic pain. At this stage it is clear that several different mechanisms are involved in neuropathic pain behavior in animals. However, the assumption that these mechanisms play a role in patients is one purpose of human research. Based on results from animal studies it is hypothesized that distinct pathophysiological mechanisms lead to specific sensory symptoms and signs in patients. Therefore, a thorough analysis of sensory symptoms and signs may reveal the underlying mechanisms that are mainly active in a particular patient. This provides a basis towards a mechanism-based treatment approach to increase therapeutic efficacy. The key method to create a detailed sensory profile of the affected painful area is quantitative sensory testing (QST) which uses different psychophysiological methods to investigate the function of small fiber afferents including nociceptive pathways as well as large fiber afferents. Interestingly, three major subtypes of distinct sensory symptom constellations can be identified in PHN which are suggested to be caused by different pathophysiological mechanisms (fig. 3).

Type I. Peripheral and Central Sensitization of Nociceptive Neurons

Abnormal nociceptor sensitization and abnormal spontaneous afferent activity has been demonstrated in many peripheral nerve injury models. Partial nerve lesion is associated with dramatic changes in the regulation of receptors and channels in damaged as well as undamaged primary afferent neurons (fig. 4). These neurons develop spontaneous activity (ectopic discharge) and an increased sensitivity to chemical, thermal and mechanical stimuli. Ectopic impulse generation following nerve injury is associated with enhanced expression and changes in the distribution of certain voltage gated sodium channels in primary afferent neurons which leads to a lowering of the action potential threshold. The expression of $Na_v1.3$ and $Na_v1.8$ protein was also induced in infected DRG following VZV infection (Fleetwood-Walker, IASP-congress).

As a consequence of periphereal nociceptor hyperactivity also dramatic secondary changes in the spinal cord dorsal horn occur (fig. 4). Partial peripheral nerve injury leads to an increase in the general excitability of spinal cord neurons. This so called central sensitization is probably due to activity in pathologically sensitized C-fibers, which sensitize spinal cord dorsal horn neurons by releasing glutamate and the neuropeptide substance P. Neuronal voltage-gated Ca-channels that are located presynaptically at the nociceptive terminals are

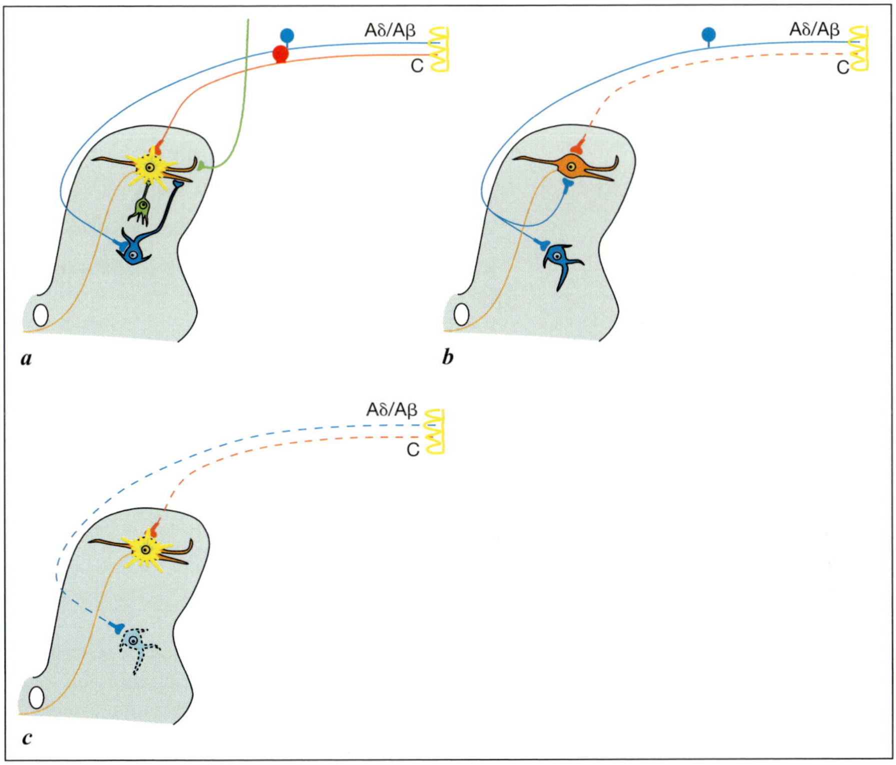

Fig. 3. Different concepts of the generation of neuropathic pain in postherpetic neuralgia. Simplified scheme. ***a*** Peripheral sensitization, central sensitization. Partial nerve lesions induce pathological activity sensitization processes in peripheral nociceptors (peripheral sensitization), leading to spontaneous burning pain, static mechanical hyperalgesia and heat hyperalgesia. This spontaneous activity in nociceptors induces secondary changes in the central sensory processing leading to spinal cord hyperexcitability (central sensitization, star in spinal cord) that causes input from mechanoreceptive Aβ-fibers (light touching) and Aδ-fibers (punctate stimuli) to be perceived as pain (dynamic and punctate mechanical allodynia). ***b*** Synaptic reorganization after C-nociceptor degeneration. Under different circumstances nociceptor function may be selectively impaired and the fibers degenerated after nerve lesion. Accordingly the synaptic contacts between central nociceptor terminals and secondary nociceptive neurons are reduced. Central terminals from intact mechanoreceptive Aβ-fibers start to sprout to form novel synaptic contacts with the 'free' central nociceptive neurons. This anatomical reorganization in the dorsal horn causes input from mechanoreceptive Aβ-fibers (light touching) to be perceived as pain (dynamic mechanical allodynia). In such patients temperature sensation are profoundly impaired in areas of severe allodynia. ***c*** Complete deafferentation. After very severe zoster infection there is a complete cutaneous deafferentation of the painful area. The dorsal root ganglion cells and the central connections of all afferents are lost in such patients and that their pain must be the result of intrinsic CNS changes, hyperactivity of second order afferent neurons.

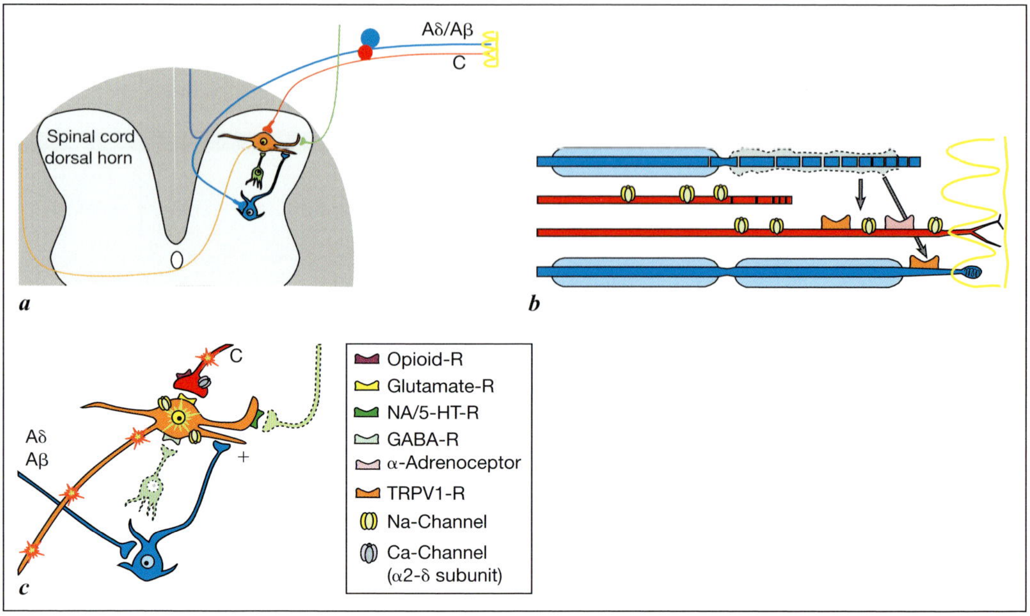

Fig. 4. Mechanisms of peripheral sensitization and central sensitization in neuropathic pain. ***a*** Schematic drawing of the primary afferent pathways and their connections in the spinal cord dorsal horn. Note that nocicepetive C-fibers (red) terminate at spinothalamic projection neurons in upper laminae (orange neuron). Non-nociceptive myelinated A-fibers project in deeper laminae. The second order projection neuron is of wide dynamic range type (WDR), i.e. it receives direct synaptic input from nociceptive terminals and also multisynaptic input from myelinated A-fibers (non-noxions information, blue neuron system). GABAergic interneurons (green neuron) normally exert inhibitory synaptic input on the WDR-neuron. Furthermore, descending modulatory systems synapse at the WDR-neuron (only the inhibitory projection, green descending terminal). ***b*** Schematic drawing of peripheral changes at primary afferent neurons after partial nerve lesion leading to peripheral sensitization. Note that some axons are damaged and degenerate (upper two) and some are still intact and connected with the peripheral end organ (skin) (lower two axons). Triggered by the lesion, Na-channels are expressed at damaged neurons. Furthermore, products such as nerve growth factor, which are associated with Wallerian degeneration and are released in the vicinity of spared fibers (arrows), trigger channel and receptor expression (Na-channels, TRPV1 receptors, adrenoreceptors) on uninjured fibers. ***c*** Spontaneous activity in C-nociceptors induces secondary changes in the central sensory processing leading to spinal cord hyperexcitability (central sensitization of second order nociceptive neurons, star in orange neuron) that causes input from mechanoreceptive A-fibers (blue neuron system, light touching and punctate stimuli) to be perceived as pain (dynamic and punctate mechanical allodynia, + = gating at synapse). Several pre- (opioid receptors, Ca-channels) and postsynaptic molecular structures (glutamate receptors, NA/5-HT receptors, GABA receptors, Na-channels) are involved in central sensitization. Inhibitory interneurons and descending modulatory control systems (green neurons) are dysfunctional after nerve lesions leading to disinhibition or faciliatation of spinal cord dorsal horn neurons and to further central sensitization. Modified from [9].

up-regulated after peripheral nerve injury and play an important role in the process of central sensitization by mediating the release of glutamate and substance P. Recently, it has been shown that VZV infection induces an upregulation of the α2δ subunit of voltage-gated calcium channel in DRG, This is of particular interest as α2δ has been proposed as one of the targets of the antiepileptic, gabapentin and pregabalin. If central sensitization is established, normally innocuous tactile stimuli become capable of activating spinal cord pain signaling neurons via Aβ-low threshold mechanoreceptors. By this mechanism light touching of the skin induces pain, i.e. mechanical allodynia.

Several clinical observations support the concept of sensitized nociceptors and central sensitization in PHN patients underlying pain and allodynia. Using QST it has been shown that about 30% of patients with PHN have severe mechanical allodynia and do not show any loss of sensory function in the affected skin. In contrast to the above histopathological data in these particular group of patients loss of neurons is minimal or absent. Accordingly, thermal sensory thresholds in their region of greatest pain are either normal or even decreased by up to 2–4°C [11, 12]. The decrease of heat pain perception thresholds (heat evoked pain, heat hyperalgesia) is a well-known phenomenon of peripheral nociceptor sensitization and is likely be associated with an up-regulation of TRPV1 receptors (fig. 4).

Using skin punch biopsy, it was shown that thermal sensitivity is directly correlated with density of cutaneous innervation in the area of most severe pain [13]. This supports the idea that, for some 30% of PHN patients pain severity is associated with relative preservation and abnormal hyperactivity (sensitized nociceptors) rather than loss of primary afferents [14, 15].

The occurrence of sensitized nociceptors is further supported by observations that topical capsaicin therapy can provide pain relief [16]. At high concentrations or with prolonged or repeated application it leads to inactivation of primary afferent nociceptors [17, 18]. Therefore, capsaicin is thought to be effective in patients with sensitized nociceptors.

Patients with sensitized nociceptors would be expected to have central sensitization as the cause of their allodynia. Since central sensitization involves the NMDA receptor, the fact that the NMDA receptor antagonist ketamine relieves pain in PHN [19] further supports the concept of peripheral and central sensitization in this condition.

In addition to their enhanced reactivity to thermal stimuli sensitized nociceptors also acquire sensitization to chemical stimuli. For example, after experimental nerve injury, surviving cutaneous afferents develop noradrenergic sensitivity, i.e. the neurons express functional adrenoreceptors at their plasma membrane (fig. 4) [9]. Furthermore, after acute infection of cultured DRG neurons with human varicella–zoster virus norepinephrine-induced Ca^{2+} influx was increased indicating adrenergic sensitivity of these neurons. These results may explain that not only mechanical lesion but also viral infection may lead to phenotypical changes of afferent neurons [20].

Clinical studies support the idea that zoster infection may trigger catecholamine sensitivity in nociceptors. In postherpetic neuralgia intracutaneous injection of adrenaline and phenylephrine increase spontaneous pain and allodynia on the affected side [21]. There are also reports that sympathetic blocks, i.v. phentolamine and topical clonidine [22] transiently relieve pain in the occasional PHN patient.

Type II. Predominant Degeneration of Nociceptive Neurons

In contrast to the patient population with sensitized nociceptors up to 60% of PHN patients show considerable signs of neuronal degeneration and loss of function within the affected tissues. Interestingly, many of these patients still suffer from severe dynamic mechanical allodynia although the function of nociceptors is diminished or absent in the same skin area.

Accordingly, Head and Campbell [3] reported extensive damage in dorsal root ganglia supplying the affected dermatomes in cases with herpes zoster, however, they made no attempt to correlate the pathology with the distribution of the pain. Subsequent pathological studies have also demonstrated loss of nerve fibers or cell bodies in the peripheral nerve, the dorsal root ganglion and also the spinal cord dorsal horn [4–7]. This is not surprising since there is viral nucleic acid in the satellite cells of dorsal root ganglia and viral reactivation produces significant local inflammation with infection and damage of primary afferents (fig. 2).

Using immunohistochemical methods to visualize nerve terminals in the skin it has been shown that cutaneous innervation is reduced in some PHN patients. Punch skin biopsies and the anti-PGP 9.5 antibody, a pan-axonal marker, in PHN patients and zoster patients without pain were used in order to quantify sensory neurites in the affected skin and compared the numbers with the homologous contralateral site [13, 23, 24]. Furthermore, a skin site distant from the shingles involvement was also analyzed. In PHN a severe dendritic loss could be demonstrated at the affected side (20% as compared with the controls). Neurite loss was more prominent in the epidermis than in the dermis. Furthermore, the PHN group also had lost half of the neurites in the contralateral epidermis whereas distant areas where unaffected.

Functional studies support the concept of degeneration of cutaneous C-nociceptors. By using these C-fiber axon reflex reactions it is possible to objectively assess cutaneous C-fiber function in the human skin. In some patients the histamine evoked axon reflex vasodilatation and flare size was impaired or abolished in skin regions with intense dynamic allodynia [25, 26]. Using quantitative thermal sensory testing to assess C- and Aδ-fibers function, some patients with acute herpes zoster [27] and some chronic PHN patients have extremely high thermal thresholds in areas with marked dynamic allodynia [21, 27, 28]. Thus, there is a subset of PHN patients with pain and *loss* of cutaneous C-nociceptor function in a region that is coextensive with allodynic skin.

The question remains which mechanisms will induce dynamic mechanical allodynia in the subgroup of patients characterized by loss of nociceptive neurons. One interesting type of central anatomical reorganization might produce mechanical allodynia. Neurons in lamina II (substantia gelatinosa) of the spinal cord dorsal horn normally receive direct input from small diameter (Aδ-, C-)fibers and respond best to noxious stimulation. Peripheral C-fiber loss results in a substantial degeneration of C-fiber terminals in lamina II. As a consequence of this loss of synaptic contacts normally made by C-fiber afferents onto pain signaling neurons in lamina II, the central terminals of Aβ-mechanoreceptive afferents which normally terminate in deeper laminae (III and IV) grow into lamina II and directly contact the deafferented cells [29]. This sprouting of Aβ-terminals depends critically on the extent of C-fiber degeneration. Such changes might be the neural correlate of the dynamic mechanical allodynia observed in some patients who have loss of C-fiber function rather than C-fiber sensitization.

This charming idea was supported by histological experiments showing specific Aβ-fiber labelling with cholera toxin B in superficial layers after nerve injury. However, recent publications question the role of Aβ-sprouting in inducing allodynia. It was shown that nerve injury induces neuroplastic changes in C-fiber nociceptors so that besides Aβ-fibers also C-fibers are labeled with cholera toxin B. In addition, large Aβ-fibers showed only very limited sprouting into inner lamina II [30, 31].

Therefore, alternative explanations for pain and allodynia in the setting of impaired cutaneous C-fiber function must be considered: (1) The nociceptive C-fibers may degenerate exclusively in the peripheral branch leaving the dorsal root ganglion soma and the central axon branch intact. The cell bodies of these injured neurons could then generate ectopic impulse activity that would maintain the pathologic central sensitization in the presence of analgesic skin. (2) In theory, since sensory testing have only studied the function of *cutaneous* C-fibers, ongoing activity that drives central sensitization might originate in intact nociceptors of deep somatic tissues (e.g. muscle, ligaments, etc.).

Type III. Complete Skin Deafferentation

Clinically a third smaller group of PHN patients can be distinguished which accounts for less than 10%. These patients have severe spontaneous pain, profound sensory loss but no evoked sensations (hyperalgesia or allodynia). In association with pain there is a complete cutaneous deafferentation of the painful area (anesthesia dolorosa). It must be assumed that the dorsal root ganglion cells and the central connections of all afferents are lost in such patients and that their pain must be the result of intrinsic CNS changes. In animal studies, following complete primary afferent loss of a spinal segment, many dorsal horn cells begin to fire spontaneously at high frequencies [32].

Mechanism-Based Treatment Approach in Postherpetic Neuralgia

Based on the described different types of PHN patients the first prospective mechanism-based treatment approach was performed using topical lidocaine [33]. All included 18 patients suffered from spontaneous burning pain and mechanical allodynia but differ concerning their cutaneous nociceptor function investigated by QST. Six patients had evidence for sensitized nociceptors in the affected dermatome. The other twelve patients demonstrated severe partial nerve injury associated with functional deafferentation including nociceptive C-fiber afferents of the affected skin area. Topical lidocaine (patch, 5%) was chosen for treatment, because it has been proven to be effective in PHN [34]. Additionally, it is generally assumed that its pain relieving effect is caused by acting on sensitized cutaneous nociceptors that have expressed voltage-gated sodium channels within the superficial layers of the skin [35]. Therefore, it was hypothesized that patients with sensitized nociceptors respond well to lidocaine. Surprisingly, topical lidocaine was more effective in patients with predominant degeneration of nociceptive neurons without inducing significant relief in patients with sensitized nociceptors. Though, the heterogeneous distribution of the patients in the two groups (6 vs. 12) weakens the statistical analyses, the striking finding that patients with nociceptor-deprived skin responded significantly to dermal lidocaine therapy cannot be ignored.

Three Distinct Sensory Profiles in PHN-Patients – Clinically Relevant?

As attractive, the PHN subtype-classification based on the nociceptor function and evoked pain types might be it should be emphasized that not all patients fit exactly into one category or the other. In a large group of PHN patients many heterogenous patterns of sensory dysfunction were detected [12]. Accordingly, detailed sensory testing in one PHN patient clearly showed areas of relative preservation in close vicinity to impaired thermal sensation, both within the affected dermatome [15]. Furthermore, the sensory patterns showed a variation over the time course of PHN.

On the other hand, several pain underlying mechanisms are probably not identified so far. For example, an explanation for the lidocaine effect in nociceptor-deprived skin might be that other intact afferents that have survived the virus infection are targeted by lidocaine after expression of sodium channels during the acute inflammatory disease process.

However, despite these limitations continuing investigations of mechanism-based treatment approaches seem to be a promising way to establish an optimal therapy with drugs that address the specific mechanisms in each patient in the future.

References

1 Portenoy RK, Duma C, Foley KM: Acute herpetic and postherpetic neuralgia: clinical review and current management. Ann Neurol 1986;20:651–664.

2 Haanpaa M, Hakkinen V, Nurmikko T: Motor involvement in acute herpes zoster. Muscle Nerve 1997;20:1433–1438.

3 Head H, Campbell AW: The pathology of herpes zoster and its bearing on sensory location. Brain 1900;23:353–523.

4 Noordenbos W: Pain. Amsterdam, Elsevier, 1959.

5 Zacks SI, Langfitt RW, Elliott FA: Herpetic neuritis: a light and electron microscopic study. Neurology 1964;14:744–750.

6 Watson CP, Morshead C, Van der Kooy D, Deck J, Evans RJ: Post-herpetic neuralgia: post-mortem analysis of a case. Pain 1988;34:129–138.

7 Watson CP, Deck JH, Morshead C, Van der Kooy D, Evans RJ: Post-herpetic neuralgia: further post-mortem studies of cases with and without pain. Pain 1991;44:105–117.

8 Baron R: Peripheral neuropathic pain: from mechanisms to symptoms. Clin J Pain 2000;16 (suppl):S12–S20.

9 Baron R: Mechanisms of disease: neuropathic pain – a clinical perspective. Nat Clin Pract Neurol 2006;2:95–106.

10 Fleetwood-Walker SM, Quinn JP, Wallace C, Blackburn-Munro G, Kelly BG, Fiskerstrand CE, Nash AA, Dalziel RG: Behavioural changes in the rat following infection with varicella-zoster virus. J Gen Virol 1999;80(pt 9):2433–2436.

11 Rowbotham MC, Fields HL: The relationship of pain, allodynia and thermal sensation in post-herpetic neuralgia. Brain 1996;119(pt 2):347–354.

12 Pappagallo M, Oaklander AL, Quatrano-Piacentini AL, Clark MR, Raja SN: Heterogenous patterns of sensory dysfunction in postherpetic neuralgia suggest multiple pathophysiologic mechanisms. Anesthesiology 2000;92:691–698.

13 Rowbotham MC, Yosipovitch G, Connolly MK, Finlay D, Forde G, Fields HL: Cutaneous innervation density in the allodynic form of postherpetic neuralgia. Neurobiol Dis 1996;3:205–214.

14 Fields HL, Rowbotham M, Baron R: Postherpetic neuralgia: irritable nociceptors and deafferentation. Neurobiol Dis 1998;5:209–227.

15 Petersen KL, Fields HL, Brennum J, Sandroni P, Rowbotham MC: Capsaicin evoked pain and allodynia in post-herpetic neuralgia. Pain 2000;88:125–133.

16 Bernstein JE, Korman NJ, Bickers DR, Dahl MV, Millikan LE: Topical capsaicin treatment of chronic postherpetic neuralgia. J Am Acad Dermatol 1989;21(pt 1):265–270.

17 Lynn B: Capsaicin: actions on nociceptive C-fibres and therapeutic potential. Pain 1990;41: 61–69.
18 Rains C, Bryson HM: Topical capsaicin. A review of its pharmacological properties and therapeutic potential in post-herpetic neuralgia, diabetic neuropathy and osteoarthritis. Drugs Aging 1995;7:317–328.
19 Eide PK, Jorum E, Stubhaug A, Bremnes J, Breivik H: Relief of post-herpetic neuralgia with the N-methyl-D-aspartic acid receptor antagonist ketamine: a double-blind, cross-over comparison with morphine and placebo. Pain 1994;58:347–354.
20 Kress M, Fickenscher H: Infection by human varicella-zoster virus confers norepinephrine sensitivity to sensory neurons from rat dorsal root ganglia. FASEB J 2001;15:1037–1043.
21 Choi B, Rowbotham MC: Effect of adrenergic receptor activation on post-herpetic neuralgia pain and sensory disturbances. Pain 1997;69:55–63.
22 Abadir AR, Kraynack BJ, Mayda J 2nd, Gintautas J: Postherpetic neuralgia: response to topical clonidine. Proc West Pharmacol Soc 1996;39:47–48.
23 Oaklander AL: The density of remaining nerve endings in human skin with and without postherpetic neuralgia after shingles. Pain 2001;92:139–145.
24 Oaklander AL: Unilateral postherpetic neuralgia is associated with bilateral sensory neuron damage. Neurology 1998;44:789–795.
25 Baron R, Saguer M: Mechanical allodynia in postherpetic neuralgia: evidence for central mechanisms depending on nociceptive C-fiber degeneration. Neurology 1995;45(suppl 8):S63–S65.
26 Baron R, Saguer M: Postherpetic neuralgia. Are C-nociceptors involved in signalling and maintenance of tactile allodynia? Brain 1993;116(pt 6):1477–1496.
27 Nurmikko T, Bowsher D: Somatosensory findings in postherpetic neuralgia. J Neurol Neurosurg Psychiatry 1990;53:135–141.
28 Nurmikko T, Wells C, Bowsher D. Sensory dysfunction in postherpetic neuralgia; in Boivie J, Hansson P, Lindblom U (eds): Touch, Temperature, and Pain in Health and Disease: Mechanisms and Assessments. Seattle, IASP Press, 1994, pp 133–141.
29 Woolf CJ, Shortland P, Coggeshall RE: Peripheral nerve injury triggers central sprouting of myelinated afferents. Nature 1992;355:75–78.
30 Santha P, Jancso G: Transganglionic transport of choleragenoid by capsaicin-sensitive C-fibre afferents to the substantia gelatinosa of the spinal dorsal horn after peripheral nerve section. Neuroscience 2003;116:621–627.
31 Bao L, Wang HF, Cai HJ, Tong YG, Jin SX, Lu YJ, Grant G, Hokfelt T, Zhang X: Peripheral axotomy induces only very limited sprouting of coarse myelinated afferents into inner lamina II of rat spinal cord. Eur J Neurosci 2002;16:175–185.
32 Lombard MC, Larabi Y: Electrophysiological study of cervical dorsal horn cells in partially deafferented rats; in Bonica JJ (ed): Advances in Pain Research and Therapy. New York, Raven Press, 1983, pp 147–154.
33 Wasner G, Kleinert A, Binder A, Schattschneider J, Baron R: Postherpetic neuralgia: topical lidocaine is effective in nociceptor-deprived skin. J Neurol 2005;252:677–686.
34 Meier T, Wasner G, Faust M, Kuntzer T, Ochsner F, Hueppe M, Bogousslavsky J, Baron R: Efficacy of lidocaine patch 5% in the treatment of focal peripheral neuropathic pain syndromes: a randomized, double-blind, placebo-controlled study. Pain 2003;106:151–158.
35 Argoff CE: New analgesics for neuropathic pain: the lidocaine patch. Clin J Pain 2000;16(suppl): S62–S66.

Prof. Dr. Med. Ralf Baron
Klinik für Neurologie, UK-SH, Campus Kiel
Sektion Neurologische Schmerzforschung und Therapie
Schittenhelmstr. 10
DE–24105 Kiel (Germany)
Tel. +49 431 597 8505, Fax +49 431 597 8530, E-Mail r.baron@neurologie.uni-kiel.de

Gross G, Doerr HW (eds): Herpes Zoster.
Monogr Virol. Basel, Karger, 2006, vol 26, pp 81–92

Varicella–Zoster Virus Infections during Pregnancy

Andreas Sauerbrei, Peter Wutzler

Institute of Virology and Antiviral Therapy, Friedrich-Schiller University, Jena, Germany

Frequency and Consequences of Varicella–Zoster Virus Infections during Pregnancy

In most industrial countries, chickenpox is a rare disease during pregnancy as more than 90% of women of child-bearing age possess virus-specific IgG class antibodies. Only 3–4% of women in Germany were found to be susceptible to primary varicella–zoster virus (VZV) infection [1]. In early reports, the average incidence of varicella in pregnant women was calculated as 0.7 per 1,000 pregnancies [2, 3], but the current rates appear to be 2–3 per 1,000 pregnancies [4].

Varicella during pregnancy may occasionally lead to serious maternal and fetal diseases (table 1). Pregnant women who contract varicella are at risk of severe pneumonia associated with life-threatening ventilatory compromise and death. The disease seems to occur more often in the third trimester [5] and must be regarded as a medical emergency. In the general population, varicella pneumonia has a mortality of 10–20% but in pregnancy it may be as high as 45%. More recent studies, however, suggest that the mortality has decreased to 10–11% for both non-pregnant and pregnant patients most likely due to the effects of antiviral therapy and better respiratory management [6].

At any stage during pregnancy, chickenpox may cause intrauterine infection. Maternal varicella resulting in viremia may transmit the virus to the fetus by either transplacental spread, or by ascending infection from lesions in the birth canal. Furthermore, direct contact or respiratory droplet can lead to infection after birth. The consequences for the infant depend on the time of maternal disease. They range from asymptomatic infection to fetal loss especially in case of severe maternal disease. Primary VZV infection during first two trimesters

Table 1. Varicella–zoster virus infections and their potential consequences during pregnancy

Maternal varicella/zoster timing	Consequences for mother/ fetus/term neonate
Varicella at any stage	Intrauterine death, neonatal or infantile zoster
Varicella during the 5th–20th (24th) weeks	Congenital varicella syndrome (risk: 2%, mortality: 30%)
Varicella at any stage, especially in third trimester	Maternal pneumonia (risk: 10–20%, mortality: 10–45%)
Varicella near term: ≥5 days before delivery	Neonatal varicella at ages 10 (–12) days (risk: 20–50%, mortality: 0%)
Varicella near term: ≤4–5 days before to 2 days after delivery	Neonatal varicella 0–4 days after birth (risk: 20–50%, mortality: 0–3%)
	Neonatal varicella 5–10 (–12) days after birth (risk: 20–50%, mortality: 20–25%)
Normal zoster at any stage	No risk for severe maternal, fetal or neonatal infections

of pregnancy may result in intrauterine infection in up to 25% of the cases [7–10]. The rate of abortion following acute varicella does not exceed the rate of abortion in pregnant women without chickenpox [7–11]. A congenital varicella syndrome (CVS) can be expected in about 12% of infected fetuses [8]. On the basis of prospective studies in Europe and North America, the incidence of embryopathy and fetopathy after maternal varicella infection in the first 20 weeks of pregnancy is estimated to be about 1–2% [9, 11]. Maternal infection near-term is associated with a substantial risk of neonatal varicella. Serious disseminated infections with visceral involvement may occur in the infant [12].

Nearly 20% of infants with intrauterine acquired VZV primary infection develop neonatal or infantile zoster, usually with uncomplicated course [13]. The disease is thought to represent reactivation of the virus after primary infection in utero. The relatively short viral latency period may be explained by the immature cell-mediated immune response in young children.

On the basis of current knowledge, zoster during pregnancy is not associated with birth defects [9, 14]. Although there are some reports of infants with congenital malformations being born to mothers with a history of zoster during early pregnancy, no case showed laboratory evidence of intrauterine infection with VZV. In addition, maternal zoster during the perinatal period does not cause problems for newborn infants [15] as the infants possess specific maternal IgG class antibodies and there is usually no longer viremic spread of VZV unless the woman is immunocompromised.

Congenital Varicella Syndrome

Clinical Manifestations

Since the first report by Laforet and Lynch [16], nearly 130 infants born with signs of CVS have been described in the English and German literature, most of them during the last 10–15 years [17]. In principal, CVS has to be expected after maternal chickenpox between the 5th and 24th gestational weeks. Nearly 80% of all cases have been observed between the 9th and 20th weeks of gestation. Before the 5th and after the 24th gestational weeks, the probability of CVS is extremely low.

The characteristic clinical findings consist of skin lesions in dermatomal distribution (fig. 1), neurological defects, eye diseases, and limb hypoplasia (table 2). Less frequent abnormalities include muscle hypoplasia, affections of the internal organs as well as gastrointestinal, genitourinary, and cardiovascular manifestations [13]. There were only small differences regarding to the dependence of symptoms on the onset of maternal chickenpox. In early infection, neurological defects and limb hypoplasia were more numerous than skin lesions and eye diseases which were dominant when maternal disease occurred later. No relationship has been reported in the literature between number of clinical features, gestational age of maternal varicella and immune response in the infant [4]. Nearly 30% of infants born with signs of CVS died during the first months of life. A follow-up report in the literature demonstrates that in spite of initially poor prognosis a good long-term outcome can occur in patients with CVS [18].

The route of fetal infection is considered to be transplacental. Ascending infection from the epithelium of the cervix uteri is also conceivable [19]. On the basis of the segmental distribution of some of the signs, it was postulated that the CVS is not the immediate consequence of intrauterine varicella, but caused by intrauterine zoster-like VZV reactivations with accompanying encephalitis [20]. In a recently published case report, a widespread non-productive VZV infection has been described in non-neural fetal tissues within 2 weeks following the onset of chickenpox in the mother [21]. Immunologic studies suggest that the fetus is not able to mount a VZV-specific cell-mediated immune response [22].

Diagnosis

Most cases of CVS have been reported on the basis of the described main clinical symptoms without laboratory evidence of intrauterine infection. However, the causal relationship between maternal varicella infection and congenital abnormalities would be most convincingly verified by detection of the virus, viral antigens or viral DNA in the infant. With the use of polymerase chain reaction (PCR) and nucleic acid hybridization assays, VZV DNA can be detected in fetal or infantile tissue samples, cerebrospinal fluid and/or amniotic

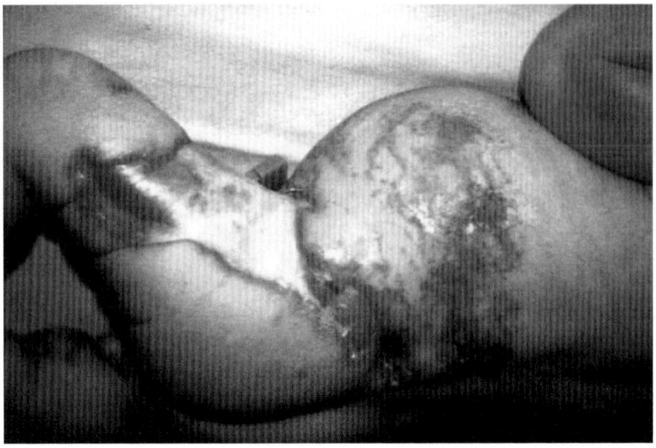

Fig. 1. Female neonate with skin lesions of the left flank and the left lower extremity after maternal varicella during the 8th–10th gestational weeks.

Table 2. Main symptoms of infants with congenital varicella syndrome cited in the literature

Symptoms	Children (n = 124)	
	n	%
Skin lesions (cicatricial scars, skin loss)	89	72
Neurological defects or diseases (cortical atrophy, spinal cord atrophy, limb paresis, seizures, microcephaly, Horner's syndrome, encephalitis, dysphagia)	77	62
Eye diseases (microphthalmia, enophthalmia, chorioretinitis, cataract, nystagmus, anisocoria, optic atrophy)	65	52
Limb hypoplasia and other skeletal anomalies	55	44
Intrauterine retardation	28	23
Gastrointestinal abnormalities	25	20
Muscle hypoplasia	24	19
Genitourinary abnormalities	15	12
Affections of internal organs	14	11
Developmental delay	13	10
Defects of the cardiovascular system	9	7
Defects of other organs	9	7

fluid [23–25]. At present, molecular biological methods should be regularly included in the diagnosis of CVS. In particular, cases presented with rare malformations or after subclinical maternal VZV infection need confirmation by virological methods, otherwise the causal relationship between maternal infection and congenital abnormalities remains doubtful [26].

Suspected prenatal infection with VZV is usually confirmed by detection of VZV-specific antibodies. Serologic diagnosis is mostly based on the persistence of VZV-specific IgG class antibodies beyond 7 months of life when maternal antibodies should normally have disappeared [27, 28]. The presence of virus-specific IgM has only been reported in about 25% of the cases with CVS [14]. Although IgM seems to be produced in small amounts by the fetus, the detection rate depends significantly on sensitivity of enzyme immunoassays, which are most frequently used for IgM detection. Unlike in cases of intrauterine rubella or cytomegalovirus infection, VZV has not been isolated in cell cultures from any infant with CVS.

To establish a relationship between maternal VZV infection and congenital anomalies of the infant, the following criteria should be used:

(1) Appearance of maternal varicella during pregnancy,

(2) Neonate or fetus with
- congenital skin lesions in dermatomal distribution and/or
- neurological defects,
- eye diseases,
- limb hypoplasia.

(3) Proof of intrauterine VZV infection by
- detection of viral DNA using PCR and/or
- presence of specific IgM/persistence of IgG beyond 7 months of age,
- appearance of zoster during early infancy.

A variety of defects and clinical symptoms described in infants with CVS may also occur in congenital infections caused by rubella virus, cytomegalovirus, herpes simplex virus, coxsackie virus or Toxoplasma gondii [29–31]. Congenital skin defects in dermatomal distribution and microphthalmia represent the cardinal symptoms of a specific genetic disorder called MIDAS (Microphthalmus, Dermal Aplasia, Sclerokornea) syndrome [32].

Prophylaxis and Treatment

For effective prophylaxis of CVS, active immunization of seronegative women before pregnancy is recommended. Varicella vaccine, as all live-attenuated vaccines, is contraindicated in pregnant women. Pregnancy has to be avoided for at least 4 weeks following vaccination. The Pregnancy Registry, managed by the Merck Research Laboratories (USA) in collaboration with the Centers for Disease Control and Prevention (USA), records women exposed to varicella vaccine during pregnancy or within 3 months before conception. Preliminary results show no hints to any birth defects related to vaccine exposure [33]. In a case report, it was documented that the varicella vaccine virus was transmitted from a vaccinated 12-month-old boy to his pregnant mother, who subsequently developed chickenpox. After an elective abortion between the 7th and 8th weeks of gestation, no virus was detected in the fetal tissue [34].

Vaccinated persons can develop mild varicella that occurs 42 days after vaccination and represents wild virus infection. These cases have been referred as to breakthrough. The rates vary between 1 and 4% per year independent of time since immunization [35]. Most breakthrough diseases are very mild, the infectivity is relatively low and there is a low or no risk for complications [36]. Therefore, the risk for CVS from breakthrough varicella can be regarded as considerably lower than that for CVS in unvaccinated women with varicella. However, since data about the risk for CVS after breakthrough varicella are not available to date, measures should be considered as in unvaccinated women who develop varicella.

Non-immune pregnant women should be advised to avoid exposure to chickenpox and zoster. VZV-specific IgG antibodies should be measured without delay in pregnant women exposed to VZV and with a negative or indeterminate history of varicella. A woman should be regarded as susceptible, if no antibodies can be detected and there is an indeterminate or unknown status of immunity. In this case, the application of varicella–zoster immune globulin (VZIG) within 72 (–96) h has been recommended [37, 38]. The prescribed dose administered intramuscularly is 125 U/10 kg of body weight, up to a maximum of 625 U [39] or 0.5 ml/kg of body weight [40]. As alternative, 1 ml/kg of body weight can be administered intravenously [40]. Although passive immunization may theoretically reduce the risk of fetal infection, there is no evidence that this prevents fetal viremia or CVS. Thus, the primary reason for VZIG is to prevent severe chickenpox and complications in the mother. If there is a definitive past history of chickenpox, it is reasonable to assume that the woman is immune to varicella.

Pregnant women, who were adequately vaccinated with 2 doses should be regarded as immune to varicella because 99% of persons become seropositive after the second dose of vaccination [41]. Thus, following exposure, routine serologic testing and administration of VZIG are not considered necessary. Furthermore, currently used enzyme immunoassays may be too insensitive to detect vaccine-induced VZV-specific IgG class antibodies [42]. On the other hand, sensitive fluorescent antibody to membrane antigen assay or tests for the determination of the cell-mediated immune response are too laborious and/or time consuming for daily routine. Nevertheless, if a vaccinated pregnant woman was tested VZV IgG-negative, she should be managed as an vaccinated seronegative pregnant woman without varicella vaccination. However, in most cases, seronegative vaccines should have acquired VZV-specific cell-mediated immunity.

Mothers with varicella during the first or second trimester should be carefully monitored since an intrauterine infection may lead to CVS. Fetal ultrasound and magnetic resonance imaging at 16–22 weeks gestational age or 5 weeks after infection can identify signs of CVS [43, 44]. Prenatal investigations for VZV DNA in placental villi, fetal blood or amniotic fluid and for VZV IgM in fetal blood are only indicated if suspicious fetal abnormalities can be seen on ultrasound or magnetic resonance imaging [43]. One should, however, be aware that the presence of VZV DNA does not necessarily correlate with fetal disease [45]. The question of how severely the fetus is affected cannot yet be answered definitely. This and the low risk of CVS should be considered in counseling women with varicella in early pregnancy. Termination of pregnancy is only indicated if there are definitive signs of serious fetal abnormalities.

To date, there are no controlled studies concerning antiviral chemotherapy in preventing CVS [46]. As the only therapeutic agent, acyclovir (10 mg/kg every 8 h intravenously for 10 days) is indicated in pregnant women. Results from the acyclovir-in-pregnancy registry do not show teratogenic effects of the drug [47]. Comparable data has been reported for the oral administration of valacyclovir [4], whereas no data about other nucleoside analogous compounds such as famciclovir and brivudin during pregnancy are available. As acyclovir is not officially approved for the treatment of pregnant women, patients should be informed about the limited information and give consent before the drug is used.

An antiviral treatment of neonates with CVS has been described in few cases [48–50]. According to clinical observations, acyclovir therapy may be helpful especially to stop the progression of eye diseases or to prevent neurological diseases after VZV reactivations.

Neonatal Varicella

Clinical Manifestations

Neonatal varicella can be expected if a mother contracts chickenpox during the last 3 weeks of pregnancy. Maternal chickenpox near term or soon after delivery may cause severe or fatal illness in the newborn. After the disease was first recognized by Hubbard [51], hundreds of cases have been reported [12]. Maternal varicella can infect the baby by (1) transplacental viremia, (2) ascending infection during birth or (3) respiratory droplet/direct contact with infectious lesions after birth. Neonatal chickenpox occurring in the first 10 (–12) days of life has to be caused by intrauterine transmission of VZV because of the incubation period of varicella. Chickenpox after the 10th (–12th) day of the neonatal period is most likely acquired by postnatal VZV infection and has a low morbidity rate [52] as most neonates are protected by maternally derived antibodies. However, premature infants younger than 28 weeks gestation or below 1,000 g birth weight are at an increased risk for severe varicella during the first 6 weeks after birth [39, 40].

The severity of intrauterine acquired neonatal chickenpox is closely related to the time of onset of maternal infection as transplacentally transmitted antibodies may reduce the severity of symptoms in the newborn. Generalized neonatal varicella leading to death is much more likely if mothers develop the varicella rash between 4 and 5 days before and 2 days after delivery [12, 53]. After maternal varicella during this period, a fatal outcome has been reported in about 20% of the cases (table 3). These infants have been exposed to maternal viremia without having acquired protecting antibodies. Furthermore, the cell-mediated immune response of the neonate is likely insufficient to retard the hematogeneous dissemination of VZV after transplacental spread [54]. A fatal outcome is thus more likely if the neonatal disease occurs between 5 and 10 days after delivery. To our knowledge, 23% of the infants reported in the literature died from a disseminated and fulminant infection (table 3; fig. 2). Neonatal varicella within the first 4 days after birth has usually been found to be comparatively mild. Fetuses exposed to VZV between 20 and 6 days before delivery may develop neonatal chickenpox however with non-fatal course. These infants get maternal antibodies and have therefore a lower risk of complications.

Neonatal chickenpox has been occasionally referred to as '*congenital varicella*' or '*neonatal varicella syndrome*'. These terms do not allow a clear differentiation from the '*congenital varicella syndrome*' caused by maternal chickenpox in the first 2 trimesters.

Diagnosis

The diagnosis of neonatal varicella is usually based on the typical clinical picture. In case of intrauterine acquired disease, the characteristic point in time and the maternal history of chickenpox during the last weeks of pregnancy have to be considered. Serological methods have been widely used to confirm the clinical diagnosis of neonatal varicella. However, the detection of virus-specific antibodies is not useful for early diagnosis. Therefore, PCR should be used as method of choice for laboratory diagnosis of VZV infection. As patient materials serve skin swabs or biopsies, liquor specimens and tissue samples. The differential diagnosis includes herpes simplex virus and enterovirus infections [31, 55].

Prophylaxis and Treatment

To reduce the mortality from neonatal chickenpox, the date of delivery may be postponed for several days to allow maternal antibodies to pass the placental barrier. However, there are only few published case reports, which describe successfully the delay of labor when neonatal varicella must be expected [56, 57].

VZIG given intravenously at a dosage of 1 ml/kg [40] or intramuscularly at a dosage of 125 U [39] or 0.5 ml/kg [40] is indicated for neonates whose mothers have signs and symptoms of varicella between 5 days before and

Table 3. Prognosis of neonatal varicella without antiviral treatment in 136 term infants cited in the literature

Day of rash onset	Cases	
	non-fatal	fatal
Mother		
≥5* (n = 57)	57 (100%)	0
4*–2** (n = 79)	65 (82%)	14 (18%)
Neonate[1]		
0–4*** (n = 35)	34 (97%)	1 (3%)
5–10*** (n = 47)	36 (77%)	11 (23%)

*Days before delivery; **days after delivery; ***days after birth.
[1]Data of 54 neonates have not been described.

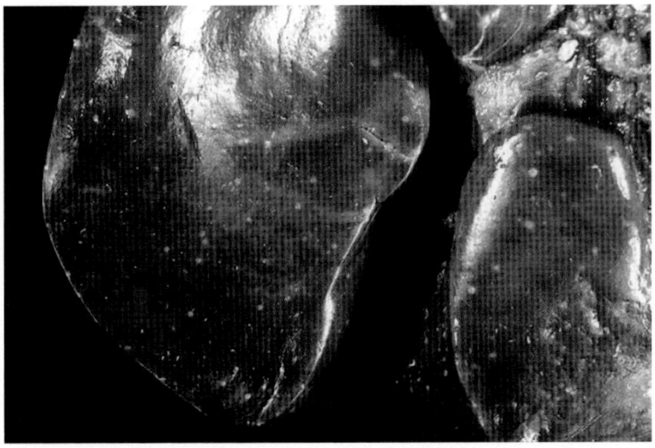

Fig. 2. Necropsy of neonate died from neonatal varicella: focal necroses on the surface of the liver.

2 days after delivery. The Department of Health Joint Committee on Vaccination and Immunisation (UK) recommends passive immunization for a period between 7 days before and 2 days after delivery [58] and the Committee on Infectious Diseases of the American Academy of Pediatrics (USA) recommends VZIG for infants whose mothers develop chickenpox between 5 days before and 3 days after delivery [8]. However, VZIG is probably not necessary for neonates whose mothers have signs of varicella >5 days before or >2 days after delivery, because those infants are not at risk of severe varicella.

Hospitalized premature infants, younger than 28 weeks gestation or below 1,000 g birth weight, who are exposed to VZV, have to receive VZIG, regardless of the maternal history of chickenpox as these infants may not have acquired maternal antibodies [39]. Following treatment, these newborns should be under surveillance in the hospital for 2 weeks, i.e. to the end of incubation period [8, 40]. When a neonate who has received VZIG is discharged home, it should be made clear to the parents that prompt hospital review should be undertaken if the baby becomes unwell or develops rash. It is generally accepted that passive immunization of the newborn can modify the clinical course of neonatal varicella, but it does not prevent the disease and although decreased, the risk of death is not eliminated [59, 60].

Therefore, acyclovir therapy should be administered promptly at a dosage of 10–15 mg/kg every 8 h intravenously for 5–7 days on suspicion of neonatal chickenpox. Prophylactic intravenous acyclovir can prevent neonatal varicella or reduce the severity of the disease markedly [61]. To date, well-controlled studies on the use of acyclovir in newborns have not been reported [46]. Mothers and newborns suffering from or being at risk of varicella have to be isolated on maternity wards.

References

1 Wutzler P, Färber I, Wagenpfeil S, Bisanz H, Tischer A: Seroprevalence of varicella-zoster virus in the German population. Vaccine 2001;20:121–124.

2 Sever J, White LR: Intrauterine viral infections. Ann Rev Med 1968;19:471–486.

3 Stagno S, Whitley RJ: Herpesvirus infections of pregnancy. Part II: Herpes simplex virus and varicella-zoster virus infections. N Engl J Med 1985;313:1327–1330.

4 Enders G, Miller E: Varicella and herpes zoster in pregnancy and the newborn; in Arvin AM, Gershon AA (eds): Varicella-zoster Virus. Virology and Clinical Management. Cambridge, University Press, 2000, pp 317–347.

5 Smego RA Jr, Asperille MO: Use of acyclovir for varicella pneumonia during pregnancy. Obstet Gynecol 1991;78:1112–1116.

6 Chandra PC, Patel H, Schiavello HJ, Briggs SL: Successful pregnancy outcome after complicated varicella pneumonia. Obstet Gynecol 1998;92:680–682.

7 Paryani SG, Arvin AM: Intrauterine infection with varicella-zoster virus after maternal varicella. N Engl J Med 1986;314:1542–1546.

8 Prober CG, Gershon AA, Grose C, McCracken GH, Nelson JD: Consensus: varicella-zoster infections in pregnancy and the perinatal period. Pediatr Infect Dis J 1990;9:865–869.

9 Enders G, Miller E, Cradock-Watson J, Bolley I, Ridehalgh M: Consequences of varicella and herpes zoster in pregnancy: prospective study of 1739 cases. Lancet 1994;343:1548–1551.

10 Liesnard C, Donner C, Brancart F, Rodesch F: Varicella in pregnancy. Lancet 1994;344:350–351.

11 Pastuszak AL, Levy M, Schick B, Zuber C, Feldkamp M, Gladstone J, Bar-Levy F, Jackson E, Donnenfeld A, Meschino W, Koren G: Outcome after maternal varicella infection in the first 20 weeks of pregnancy. N Engl J Med 1994;330:901–905.

12 Sauerbrei A, Wutzler P: Neonatal varicella. J Perinatol 2001;21:545–549.

13 Sauerbrei A, Wutzler P: Das fetale Varizellensyndrom. Monatsschr Kinderheilkd 2003;151: 209–213.

14 Sauerbrei A, Wutzler P: The congenital varicella syndrome. J Perinatol 2000;20:548–554.

15 Miller E, Cradock-Watson JE, Ridehalgh MK: Outcome in newborn babies given anti-varicella-zoster immunoglobulin after perinatal maternal infection with varicella-zoster virus. Lancet 1989;2:371–373.

16 Laforet EG, Lynch CJ Jr: Multiple congenital dafects following maternal varicella. N Engl J Med 1947;236:534–537.

17 Sauerbrei A, Wutzler P: Varicella-zoster virus infections during pregnancy: epidemiology, clinical symptoms, diagnosis, prevention and therapy. Curr Pediatr Rev 2005;1:205–216.

18 Schulze A, Dietzsch HJ: The natural history of varicella embryopathy: a 25-year follow-up. J Pediatr 2000;137:871–874.

19 Birch CJ, Druce JD, Catton MC, MacGregor L, Read T: Detection of varicella-zoster virus in genital specimens using a multiplex polymerase chain reaction. Sex Transm Infect 2003;79:298–300.

20 Higa K, Dan K, Manabe H: Varicella-zoster virus infections during pregnancy: hypothesis concerning the mechanisms of congenital malformation. Obstet Gynecol 1987;69:214–221.

21 Nikkels A, Delbecque K, Pierard GE, Wienkötter P, Schalasta G, Enders M: Distribution of varicella-zoster virus DNA and gene products in tissues of a first-trimester varicella-infected fetus. J Infect Dis 2005;191:540–545.

22 Grose C: Congenital varicella-zoster virus infection and the failure to establish virus-specific cell-mediated immunity. Mol Biol Med 1989;6:453–462.

23 Scharf A, Scherr O, Enders G, Helftenbein E: Virus detection in the fetal tissue of a premature delivery with congenital varicella syndrome: a case report. J Perinatol Med 1990;18: 317–322.

24 Michie CA, Acolet D, Charlton R, Stevens JP, Happerfield LC, Bobrow LG, Kangro H, Gau G, Modi N: Varicella-zoster contracted in the second trimester of pregnancy. Pediatr Infect Dis J 1992;11:1050–1053.

25 Sauerbrei A, Müller D, Eichhorn U, Wutzler P: Detection of varicella-zoster virus in congenital varicella syndrome: a case report. Obstet Gynecol 1996;88:687–689.

26 Al-Katawee YA, Al-Hassoun YA, Taha MN, Al-Moslem K: Congenital varicella-zoster virus infection. A rare case of severe brain and ocular malformations without limb or cutaneous involvement in a newborn after maternal subclinical infection. Saudi Med J 2005;26:869–871.

27 Gershon AA, Raker R, Steinberg S, Topf-Olstein B, Drusin LM: Antibody to varicella-zoster virus in parturient women and their offspring during the first year of life. Pediatrics 1976;58:692–696.

28 Akisu M, Yalaz M, Aksu G, Arslanoglu S, Genel F, Kutukculer N, Kultursay N: Maternally acquired varicella-zoster virus antibodies disappear at 6 months of age in prematurely born children. Panminerva Med 2003;45:155–156.

29 Koskimies O, Lapinleimu K, Saxén L: Infections and other maternal factors as risk indicators for congenital malformations: a case-control study with paired serum samples. Pediatrics 1978;61: 832–837.

30 Johansson AB, Rassart A, Blum D, Van Beers D, Liesnard C: Lower-limb hypoplasia due to intrauterine infection with herpes simplex virus type 2: possible confusion with intrauterine varicella-zoster syndrome. Clin Infect Dis 2004;38:e57–e62.

31 Sauerbrei A, Glück B, Jung K, Bittrich H, Wutzler P: Congenital skin lesions caused by intrauterine coxsackievirus B3 infection: a case report. Infection 2000;28:326–328.

32 Spranger S, Stute H, Blankenagel A, Jauch A, Hager D, Tariverdian G: MIDAS-Syndrom – Eine X-chromosomale Erkrankung. Differenzialdiagnose zum kongenitalen Varizellensyndrom. Monatsschr Kinderheilkd 1998;146:761–765.

33 Shields KE, Galil K, Seward J, Sharrar RG, Cordero JF, Slater E: Varicella vaccine exposure during pregnancy: data from the first 5 years of the pregnancy registry. Obstet Gynecol 2001;98: 14–19.

34 Salzman MB, Sharrar RG, Steinberg S, LaRussa P: Transmission of varicella-vaccine virus from a healthy 12-month-old child to his pregnant mother. J Pediatr 1997;131:151–154.

35 Watson B: Varicella: a vaccine preventable disease? J Infect 2002;44:220–225.

36 Vázquez M, Shapiro ED: Varicella vaccine and infection with varicella-zoster virus. N Engl J Med 2005;352:439–440.

37 Royal College of Obstetricians and Gynecologists: Guideline No. 13. Chickenpox in Pregnancy. 2001.

38 Robert Koch-Institut: Empfehlungen der Ständigen Impfkommission (STIKO) am Robert Koch Institut/Stand: Juli 2004. Epidemiol Bull 2004;30:235–250.
39 Advisory Committee on Immunization Practices (ACIP): Prevention of varicella. MMWR 1996;45:1–36.
40 Deutsche Gesellschaft für Pädiatrische Infektiologie: Handbuch 2003: Infektionen im Kindesalter. München, Futuramed, 2003, pp 732–739.
41 Kuter BJ, Ngai A, Patterson CM, Staehle BO, Cho I, Matthews H, Provost PJ, White CJ: Safety, tolerability, and immunogenicity of two regimens of Oka/Merck varicella vaccine (Varivax®) in healthy adolescents and adults. Vaccine 1995;13:967–972.
42 Sauerbrei A, Wutzler P: Labordiagnostik der Varizellen. Kinderärztl Prax 2004, Sonderheft Impfen, pp 18–21.
43 Pretorius DH, Hayward I, Jones KL, Stamm E: Sonographic evaluation of pregnancies with maternal varicella infection. J Ultrasound Med 1992;11:459–463.
44 Verstraelen H, Vanzieleghem B, Deroort P, Vanhaesebrouck P, Temmerman M: Prenatal ultrasound and magnetic resonance imaging in fetal varicella syndrome: correlation with pathology findings. Prenat Diagn 2003;23:705–709.
45 Mouly F, Mirlesse V, Meritet JF, Rozenberg F, Poissonier MH, Lebon P, Daffos F: Prenatal diagnosis of fetal varicella-zoster virus infection with polymerase chain reaction of amniotic fluid in 107 cases. Am J Obstet Gynecol 1997;177:894–898.
46 Tan MP, Koren G. Chickenpox in pregnancy: revisited. Reprod Toxicol 2006;21:410–420, DOI: 10.1016/j.reprotox.2005.04.011.
47 Andrews EB: Acyclovir does not increase risk in pregnancy. Am Pharm 1994;NS34:6.
48 Wutzler P, Sauerbrei A, Scholz H, Müller D, Wiedersberg H: Varicella-Zoster-Virusinfektionen in der Schwangerschaft. Paediatr Prax 1990–1991;41:213–224.
49 Sauerbrei A, Pawlak J, Luger C, Wutzler P: Intracerebral VZV reactivation in congenital varicella syndrome? A case report. Dev Med Child Neurol 2003;45:837–840.
50 Schulze-Oechtering F, Roth B, Enders G, Grosser R: Kongenitales Varizellensyndrom – besteht eine Infektionsgefahr für die Umgebung? Dtsch Med Wochenschr 2004;208:25–28.
51 Hubbard TW: Varicella occurring in an infant twenty four hours after birth. Br Med J 1878;1:822.
52 Heidl M: Varicella-Zoster-Virus-Infektion in der Schwangerschaft, beim Neugeborenen und jungen Säugling. Z Klin Med 1985;40:245–250.
53 Sterner G, Forsgren M, Enocksson E, Grandien M, Granström G: Varicella-zoster infections in late pregnancy. Scand J Infect Dis 1990;71:30–35.
54 Baba K, Yabuuchi H, Takahashi M, Ogra PL: Immunologic and epidemiologic aspects of varicella infection acqired during infancy and early childhood. J Pediatr 1982;100:881–885.
55 Gershon AA: Varicella-zoster virus; in Feigin RD, Cherry JD (eds): Textbook of Pediatric Infectious Diseases, ed 4. Philadelphia, W. B. Saunders Company, 1998, pp 1769–1777.
56 Paulman PM, McLellan R: Varicella during pregnancy: the timing of effective treatment. J Am Board Fam Pract 1990;3:121–123.
57 Zieger W, Friese K, Weigel M, Becker KP, Melchert F: Varizellen-Infektion am Geburtstermin. Z Geburtshilfe Perinatol 1994;198:134–137.
58 Miller E: Varicella-zoster virus infection in pregnancy. Arch Dis Child 1994;70:F157–F158.
59 Holland P, Isaacs D, Moxon ER: Fatal neonatal varicella infection. Lancet 1986;2:1156.
60 King SM, Gorensek M, Ford-Jones EL, Read SE: Fatal varicella-zoster infection in a newborn treated with varicella-zoster immunoglobulin. Pediatr Infect Dis 1986;5:588–589.
61 Carter PE, Duffty P, Lloyd DJ: Neonatal varicella infection Lancet 1986;2:1459–1460.

Dr. Andreas Sauerbrei
Institute of Virology and Antiviral Therapy
Friedrich-Schiller University Jena
Hans-Knöll-Strasse 2
DE–07745 Jena (Germany)
Tel. +49 3641 657300, Fax +49 3641 657301, E-Mail Andreas.Sauerbrei@med.uni-jena.de

Gross G, Doerr HW (eds): Herpes Zoster.
Monogr Virol. Basel, Karger, 2006, vol 26, pp 93–106

Herpes Zoster in the Immunocompromised Host

Helmut Schöfer

Department of Dermatology, University Hospital, J.W. Goethe University, Frankfurt, Germany

After primary chickenpox infection and varicella–zoster virus (VZV) viraemia, which usually occurs during childhood (with or without a varicella exanthem), VZV persists as a residual latent infection in the dorsal root ganglia. At the age of 13–14 years, endemic VZV infection in Germany exceeds 90%. The latent infection is controlled by cellular immune functions (specific $CD4^+$-T-lymphocytes) and neutralising antibodies [1]. Although the pathomechanisms are not yet understood in detail, exogenic reinfection and/or immunological dysbalance, due to the decrease of T-cell related cellular immune functions and the loss of neutralising antibodies, may lead to a dermatomal reactivation in patients carrying the latent VZV infection. Clinically, this results in the manifestation of herpes zoster. Triggers for reactivation may be:

In patients >50 years:

- Age related (physiological) deficiency of cellular immune functions: Loss of $CD4^+$ T-lymphocytes by numbers ($CD4^+$-count ↓) and loss of T-cell activity.
- Therapy with immunosuppressives [2].
- Chemotherapy.
- Severe consumptive diseases.

In patients <50 years:

- Incomplete immunologic response to prenatal chickenpox infection.
- Incomplete immunologic response to chickenpox infection in the first year of life.
- Atopic diathesis.
- HIV-infection (CD4-count ↓).
- Therapy with immunosuppressives.
- Severe consumptive diseases.

- Distress (emotional, physical).
- Severe insolation (UV-irradiation).
- Pregnancy (?).
- In HIV+ patients: Immune reconstitution, following highly active anti-retroviral therapy (HAART).

Representing a disease of the elderly, the occurrence of herpes zoster in patients younger than 50 years of age is suspicious of any kind of cellular immunodeficiency. In times prior to the HIV-epidemic, malignant systemic diseases, like leukaemia or Hodgkin's disease were most likely to be the cause of shingles in young patients (<50 years of age). Since the early eighties of the 20th century the HIV-infection became more and more relevant and nowadays it is the most important disease to be excluded in young zoster patients [3].

Herpes Zoster in HIV-Infected Patients

Epidemiology

Only 3 years after the first descriptions of AIDS as a new disease Mathur-Wagh and Mildvan [4] reported, that herpes zoster is very common in the pre-AIDS stage of the HIV-disease, which was named 'lymphadenopathy syndrome' at that time. 24% of these patients had a herpes zoster manifestation prior to full-blown AIDS, which was diagnosed exclusively by clinical and immunological parameters until HIV was discovered (1983) and first antibody tests had been established (1984). In 1986, Friedman-Kien et al. [5] pointed out, that herpes zoster is an early clinical sign of the development of AIDS. He calculated a 7-fold risk for HIV-infected patients to suffer from herpes zoster compared to non-HIV-infected adults. Later Buchbinder et al. [6] showed that the age-adjusted overall relative risk of herpes zoster among HIV-infected homosexuals was 16.9 compared to non-HIV-infected homosexual men and that herpes zoster might occur during all stages of HIV disease. The herpes zoster incidence ranges from 2.9 to 5.1 episodes per 1,000 person years [6–9]. In 1986, the Centres for Disease Control (CDC, Atlanta) included shingles involving more than one dermatome in the symptomatic phase of the HIV disease and AIDS classification (B symptoms).

Today it is well-established that herpes zoster in patients less than 50 years of age, is the most striking marker disease of HIV-infection. Marker diseases are defined as illnesses that are attracting the physician's attention to the possibility of an underlying acquired immunodeficiency, i.e. HIV-infection by their mere appearance or special clinical features [10–12]. Among 188 patients of the Department of Dermatology, Frankfurt/M University Hospital, in which skin disease had led to the diagnosis of HIV-infection, herpes zoster was the presenting marker disease in 25 cases (unpublished data, Schöfer).

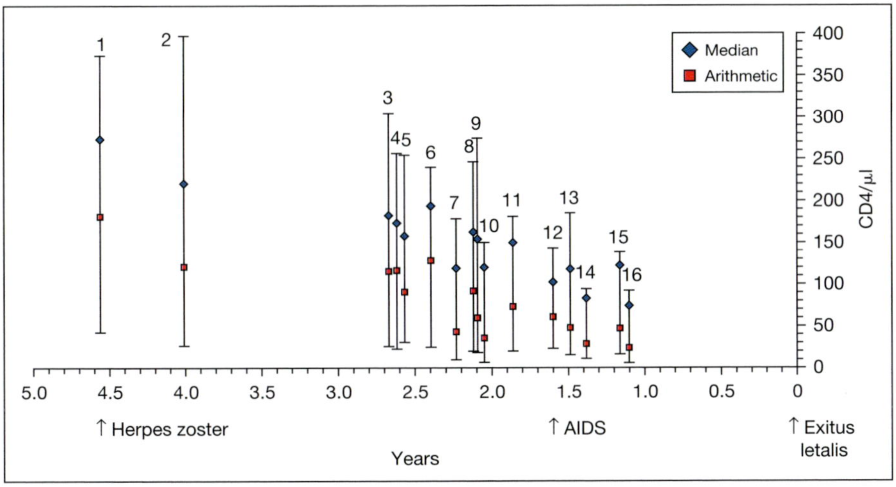

Fig. 1. Manifestation of herpes zoster during the course of HIV disease in 330 patients who died by AIDS. Median and arithmetic CD4-counts/µl (with 25 and 75 percentile) at the time of Herpes zoster manifestation (S. Reinmöller, FFM 2000). 1: Herpes zoster, 2: genital warts, 3: dermatophytosis, 4: pruritus, 5: seborrheic dermatitis, 6: folliculitis, 7: herpes labialis, 8: oral hairy leukoplakia, 9: oral candidiasis, 10: abscess, furunculosis, 11: xerosis, 12: AIDS, 13: Kaposi's sarcoma, 14: herpes genitoanalis, 15: drug eruptions, 16: mollusca contagiosa.

In many patients with HIV disease herpes zoster occurs prior to any other opportunistic infections. In patients who never had received any antiretroviral therapy, herpes zoster occurred on an average 3 years prior to any AIDS-defining illness and 4.56 years before these patients died by AIDS. The median CD4-cell count at the time of zoster manifestation was relatively high (271 $CD4^+$-cells/µl) compared to other opportunistic infections, that do occur at CD4-cell counts <100/µl (fig. 1) in common. Among 2,149 HIV-infected patients seen at the Department of Dermatology, Frankfurt/M University Hospital (1982–2000) 369 (17.2%) developed herpes zoster. The frequency among males (332/1,919; 17.3%) and females (37/230; 16.1%) was similar (fig. 2). Thirty-eight of 369 herpes zoster patients (10.3%) had one or more relapses. In other cohorts of German HIV+ patients herpes zoster was diagnosed in 20.6% [13] and 14% [14].

Comorbidity with Kaposi's Sarcoma

In 1986, Friedman-Kien et al. [5] presumed that the manifestation of shingles seems to facilitate the occurrence of Kaposi's sarcoma. This is in accordance with our epidemiological data. Every 4th HIV+ patient (25.2%) with

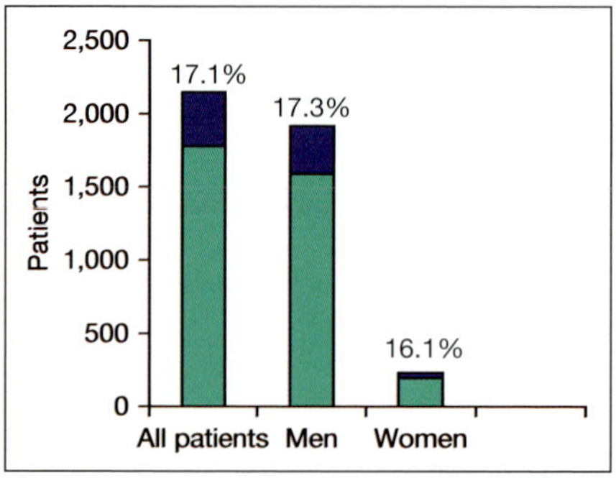

Fig. 2. Herpes zoster in 2,149 HIV-infected patients. Blue = HIV+ with herpes zoster. Green = HIV+ without herpes zoster.

herpes zoster suffered from Kaposi's sarcoma (with or after herpes zoster), whereas Kaposi's sarcoma occurred in 21.5% of all 2,149 HIV-infected patients at our clinic. A possible explanation for this comorbidity could be the fact, that both diseases are due to viruses from the herpes virus family (VZV and HHV-8). Within this family of eight pathogen viruses, co-infections with two or more viruses, affecting the same type of human cells, have been observed. In addition, it is well-known that an infection with one of these viruses (e.g. HSV-1) has an influence (±) on the transmission rates of other viruses (e.g. HSV-2) of the same family.

Clinical Features

As a rule, herpes zoster is easily diagnosed ('eye-catching diagnosis') by clinical criteria (i.e. localised prodromal and acute pain, grouped vesicles developing unilaterally with or without erythema, localised to the area of one or two dermatomes, see figs. 3, 4). In the majority of HIV-infected and other immunodeficient patients (>90%) this typical clinical aspect of herpes zoster segmentalis (figs. 3, 4) is observed [15]. But in a few cases of a so far unpublished study at the University of Frankfurt, Germany, atypical distribution, involvement of >2 dermatomes (5/369) or herpes zoster generalisatus with a varicella-like dissemination (4/369), severe haemorrhage (2/369), persistent ulcerations over months (3/369) or hyperkeratotic lesions (2/369) were observed [16]. More than 10% of all herpes zoster cases were localised at the head and neck area. Probably due to the relatively low age of our HIV-infected patients (mean age 34.7 ± 10.9 years) severe post-herpetic neuralgia, which is a feared complication in the elderly, was rarely observed (3/369; 0.8%). Herpes

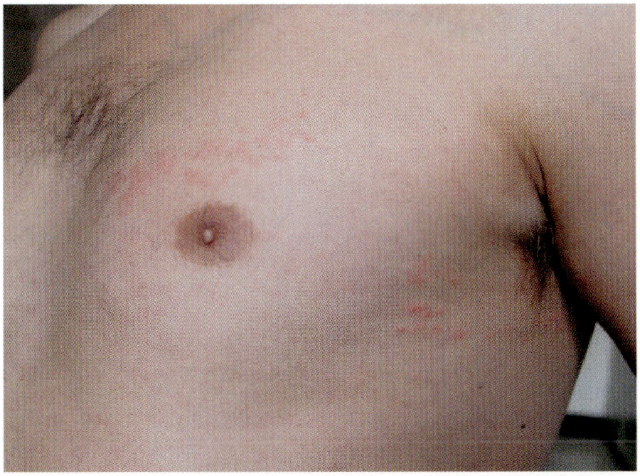

Fig. 3. RM, 34 years, acute chest pain on the left side, groups of vesicles in the left thoracic nerve segment Th 3.

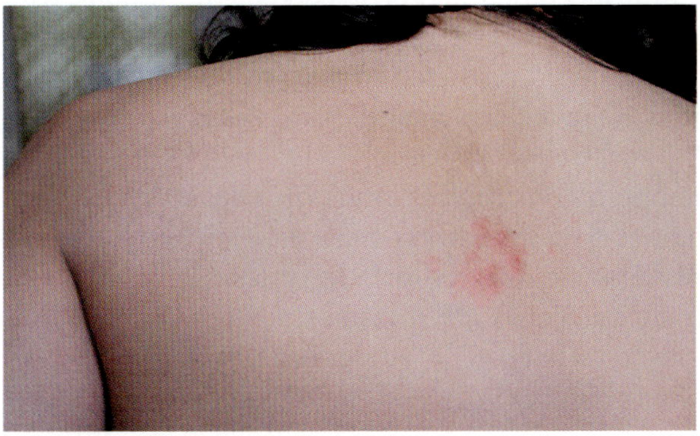

Fig. 4. Group of vesicles located at the patients back (left thoracic nerve segment Th 3). Early herpes zoster thoracalis in HIV-infection.

zoster without skin lesions (herpes zoster sine herpete) was not diagnosed in these patients (table 1). In a cohort of African herpes zoster patients the presence of systemic symptoms such as fever and weakness correlated significantly with the presence of HIV-infection [17].

Concerning skin involvement, the distribution pattern (two and more dermatomes involved, herpes zoster duplex or generalisation), a tendency to relapse or even to persist and the severity of cutaneous lesions (haemorrhagic, ulcerating, necrotising, hyperkeratotic, scarring) are of special interest (fig. 5). A higher

Table 1. Herpes zoster in 369 HIV-infected patients: clinical aspects (Schöfer, unpublished data)

Typical herpes zoster segmentalis	>90%
Involvement >2 dermatomes	1.4%
Herpes zoster generalisatus	1.1%
Zoster with persisting ulcers	0.8%
Severe post-herpetic neuralgia	0.8%
Herpes zoster hemorrhagicus	0.5%
Zoster with hyperkeratotic lesions	0.5%

frequency of disseminated herpes zoster lesions in immunocompromised patients was already reported in the pre-AIDS era by Weber and Pelecchia [18]. Dissemination rates are increased for cutaneous as well as for the very rare visceral dissemination (herpes zoster pneumonia, VZV hepatitis, VZV meningoencephalitis etc.). It was also demonstrated, that virus shedding persists 2 days longer (7.0 vs. 5.3 days) in the immunocompromised host compared to immunocompetend patients [5]. Painful ulcerating and/or hyperkeratotic lesions with persistance of viral replication for months have been rarely described. Such lesions are clinically very unusual and restricted to sites of the initial herpes zoster eruptions. Necrotic and ecthymatous shingles were described by Alessi et al. [19], Gilson et al. [20] and Hoppenjans et al. [21]. Figure 6 shows one of two cases with hyperkeratotic lesions seen at Frankfurt University Hospital [16]. Slightly different lesions with purple hyperkeratosis, peripheral erythro-oedematous halo and hyperkeratotic blisters on the soles were published by Agusto et al. [22]. Chronic verrucous VZV lesions are very rare in HIV negative patients. Recently, Jeyaratnam et al. [23] described disseminated verrucous rashes in a renal transplant recipient.

Other complications reported in literature to occur frequently in HIV-infected patients are explained below.

Herpes Zoster Ophthalmicus

Severe eye involvement was reported from African AIDS patients with herpes zoster ophthalmicus in Malawi [24]. Forty-one percent of these patients had corneal perforation. Necrotising retinitis, another variant of severe eye involvement, was diagnosed in 17% of 29 immunodeficient HIV+ patients with herpes zoster ophthalmicus in Miami by Sellitti et al. [25]. It seems possible to avoid these severe complications by early antiviral treatment: Margolis et al. [26] found sight-threatening eye involvement to be a rare condition in 48 HIV-infected with herpes zoster ophthalmicus who had early acyclovir treatment.

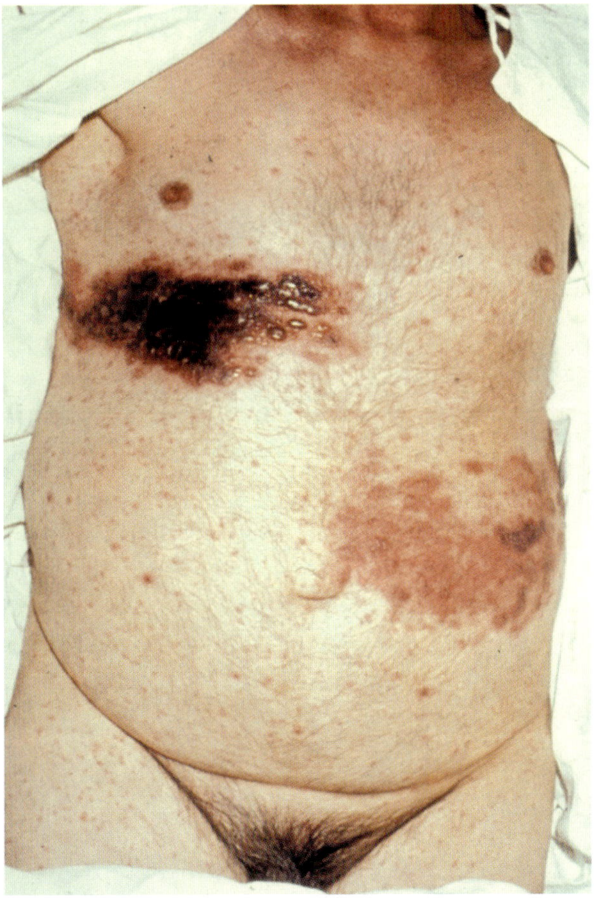

Fig. 5. Herpes zoster duplex with generalisation in a patient suffering from chronic lympathic leukaemia.

Most of the patients with VZV infection of the first division of the trigeminal nerve had no or mild eye involvement. Necrotising retinitis was diagnosed in 2/48 patients (4%), chronic infectious pseudodendritic keratitis in 2/48 (4%) and another 3/48 patients (6%) had iritis with elevations in intraocular pressure.

Neurological Complications

A higher rate of neurological complications including aseptic meningitis, myelitis, encephalitis, facial nerve palsy and radiculitis has been also reported in HIV-infected and otherwise immunodeficient patients (see chapter: 'Postherpetic

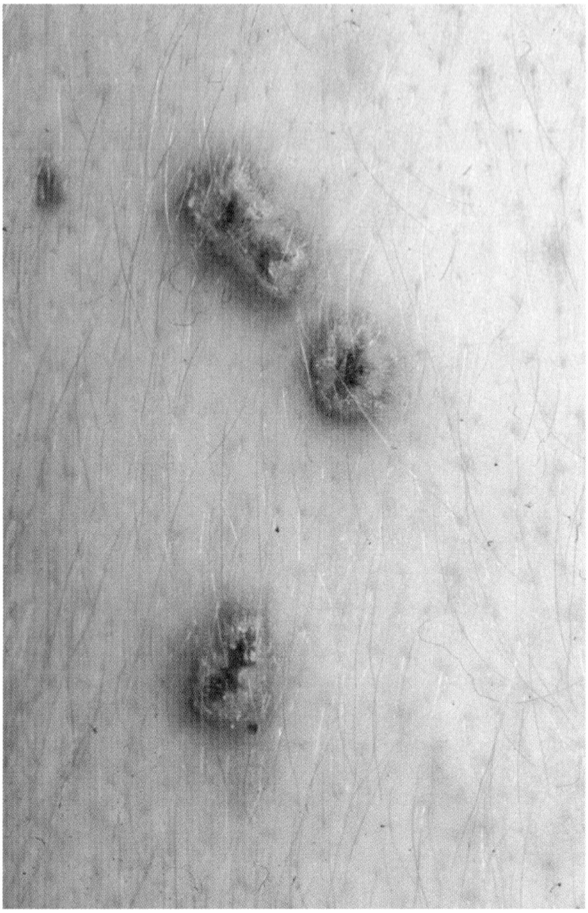

Fig. 6. Hyperkeratotic lesions of chronic herpes zoster infection (5 months).

Neuralgia and Other Neurologic Complications' by R. Baron, pp 69–80), but some authors [27, 28] demonstrated, that immunodeficiency does not increase the risk of post-herpetic neuralgia. The incidence of this complication seems to be related above all to the patient's age.

Herpes Zoster, HIV and the Immune Reconstitution Syndrome

Severe immunodeficiency ($CD4^+$-T-lymphocytes $<200/\mu l$), as observed in untreated AIDS patients, seems to hinder the clinical manifestation of herpes zoster. However, in cases where HAART had been introduced successfully ($CD4^+$-T-lymphocytes ↑, HIV viral load ↓) and had led to a relevant immune reconstitution, herpes zoster was reported to occur more frequently [29–33]. This was also observed in patients, who switched from a rather ineffective antiretroviral monotherapy to HAART [29]. In these patients herpes zoster occurred on an average 16.6 weeks after HAART had been started, but half of the patients had herpes zoster already within the first 4 weeks. Domingo et al. [30] found that shingles in connection with the immune reconstitution syndrome occurred especially in those patients who had an substantial increase of $CD8^+$-T-lymphocytes after HAART had been started. Only a few cases of zoster in the course of immune reconstitution were reported to be recurrences.

In contrast to the data published in literature, we identified only one of 94 HIV+ herpes zoster patients, who had a herpes zoster manifestation in direct association (within 3 months after the start of HAART) with immune reconstitution (unpublished data: Department of Dermatology, University Hospital Frankfurt/M, Germany 1995–2003).

Herpes Zoster and the Progression of HIV Disease Towards AIDS

Several longitudinal investigations that tried to answer the question, whether herpes zoster in HIV-infected patients increases the risk of AIDS manifestation (independent from CD4-cell counts) did not show any significant influence on the outcome of HIV-infection and survival [6, 34–37].

Herpes Zoster in Organ Transplant Recipients

It is now more than 35 years ago, that the first kidney was transplanted. Immunosuppressive regimens have changed several times since those pioneer days, but still there are substantial problems to prevent abortion of the transplanted organ. Iatrogenic immunosuppression with corticosteroids, cyclosporin A, azathioprin or mycophenolat mofetil favours viral infections. Severe cytomegalovirus and recurrent herpes simplex virus infections are frequent and alarming complications. Primary VZV infections harbour the danger of severe chickenpox with life threatening organ manifestations like VZV pneumonia, encephalitis and disseminated intravascular coagulopathy. Herpes zoster is diagnosed in 8–30% of all transplant recipients within the first 6 months of immunosuppression. Highest rates (50% within 12 months) occur in bone marrow recipients who did not have an acyclovir prophylaxis. Postherpetic neuralgia and a prolonged duration of skin disease is reported in organ transplant recipients [38]. Gourishankar et al. [39] analysed 869 recipients of solid organs for the incidence and clinical features of herpes zoster. An overall incidence of 8.6% was found. Herpes zoster was almost twice as frequent in heart and lung recipients (15.1–16.8%) compared to liver and kidney recipients (5.7–7.4%) and started 9 months (median time) after transplantation. The female gender and immunosuppressive therapy with mycophenolate mofetil were found to be independent risk factors in these patients [40]. Cutaneous scarring (18.7%) and post-herpetic neuralgia (42.7%) were frequent complications [39], whereas verrucous lesions are rare in transplant recipients [23].

In cases where a transplantation can be planned in advance, VZV vaccination is substantial in VZV negative patients and might be helpful as a booster vaccination in the elderly patient. If immunosuppression has started, vaccination with attenuated virus is no longer allowed and specific hyperimmunoglobulins have to be considered instead. In addition, a prolonged CMV prophylaxis seems to be also effective to reduce the risk of herpes zoster [39].

Herpes Zoster in Patients with Malignancies and Chemotherapy

Especially patients with malignant lymphomas (Hodgkin's disease as well as non-Hodgkin lymphoma) or leukaemia are endangered to suffer from herpes zoster due to the disease related immunodeficiency or due to chemotherapy [41–43] or bone marrow transplantation. Clinical course and manifestations do not differ much from shingles in HIV-infected patients, but cutaneous dissemination is observed in 9–16% of all cases associated with malignancies and in 6.7–25.9% in patients with Hodgkin's disease [44]. Some cases of visceral involvement masked as abdominal pain or acute abdomen have been reported in patients after bone marrow transplantation [45–47]. The overall relative risk for shingles in tumour patients (age adapted) was calculated 5.0. Guess et al. [48] reviewed 173 cases of herpes zoster in childhood and found a 122-fold risk of herpes zoster for children with acute lymphocytic leukaemia. Shingles in these children caused very few complications and not a single case of postherpetic neuralgia. After bone marrow and stem cell transplantation acyclovir or valaciclovir prophylaxis is recommended [49, 50].

It is not reasonable to screen for occult tumours in all herpes zoster segmentalis patients. The rate of occult tumours found in otherwise asymptomatic herpes zoster patients is less than 5% of the overall relative risk 1.1 [51]. In patients with disseminated zoster malignancies are found in a higher rate and tumour screening is always recommended [52].

Therapy

Only acyclovir and famciclovir (in patients older than 25 years) are registered pharmaceuticals for the treatment of herpes zoster in the immunocompromised host [53, 54]. Recommended doses are depending on the grade of immunodeficiency (with the absolute CD4-cell count as a surrogate marker) and the severity of the disease:

In patients with $CD4^+$-T-lymphocytes $\geq 200/\mu l$ and segmental zoster: Acyclovir 5–7.5 mg/kg BW every 8 h i.v., Famciclovir 3×250 mg/day perorally.

In patients with $CD4^+$-T-lymphocytes $<200/\mu l$ and/or clinical complications (haemorrhage, dissemination, neurological symptoms etc.): 10 mg/kg BW every 8 h i.v.

In case of acyclovir resistance: Foscarnet (60 mg/kg BW BID or 40 mg/kg BW TID i.v.) may be tried as an alternative. Foscarnet efficacy is independent from viral thymidinkinase.

Oral antivirals like valacyclovir or brivudin have been effective in single cases, but are not registered for the immunodeficient host. Recently, Breton et al. [55] reported the failure of valacyclovir in a HIV-infected patient with only moderate immunodeficiency.

Conclusion

As a rule, herpes zoster in the immunocompromised host is a frequent finding, but in general (>90% of all cases) as long as it is treated early and effectively there is no difference clinically from herpes zoster in the elderly (herpes zoster segmentalis). Although rare, severe courses and a wide spectrum of clinical variants and complications as well as postherpetic neuralgia are seen occasionally and cause diagnostic and therapeutic problems. Relapsing and persisting shingles with ulcerating or hyperkeratotic lesions are exclusively seen in the immunodeficient host, where virus replication cannot be controlled by immunological surveillance and acyclovir resistant strains might develop under antiviral treatment. Therapy is linked to the severity of cellular immune deficiency and to clinical complications. A new finding is the manifestation of herpes zoster as a part of the immune reconstitution syndrome in HIV-infected patients under HAART. This complication of HAART usually occurs during the first 3 months of effective treatment and is accompanied by a marked increase of $CD8^+$-T-lymphocytes. In addition, herpes zoster is an important clinical marker disease for acquired immunodeficiency. It occurs frequently as the first clinical symptom of HIV-infection, but can also be diagnosed in all other stages of HIV disease. Therefore, herpes zoster in young adults (<50 years of age) is an important indication for anti-HIV antibody testing.

References

1 Weksler ME: Immune senescence. Ann Neurol 1994;35:35–37.

2 Spencer ES, Anderson HK: Clinical evident, non-terminal infections with herpes viruses and the wart virus in immunosuppressed renal allograft recipients. Br Med J 1970;1:251–254.

3 Schöfer H, Oswald A, Allwinn R: Zoster bei 223 Patienten unter 50 Jahren – eine retrospektive Vergleichsanalyse von Patienten mit und ohne HIV-Infektion. J Dtsch Dermatol Ges 2003; 1(suppl 1):S85.

4 Mathur-Wagh U, Mildvan D: Prodromal syndromes in AIDS. Ann N Y Acad Sci 1984;437: 184–191.

5 Friedman-Kien AE, Lafleur FL, Gendler E, Hennessey NP, Montagna R, Halbert S, Rubinstein P, Krasinski K, Zang E, Poiesz B: Herpes zoster: a possible early clinical sign for development of acquired immunodeficiency syndrome in high-risk individuals. J Am Acad Dermatol 1986;14: 1023–1028.

6 Buchbinder SP, Katz MH, Hessol NA, Liu JY, O'Malley PM, Underwood R, Holmberg SD: Herpes zoster and human immunodeficiency virus infection. J Infect Dis 1992;166:1153–1156.

7 Glesby MJ, Moore RD, Chaisson RE: Clinical spectrum of herpes zoster in adults infected with human immunodeficiency virus. Clin Infect Dis 1995;21:370–375.

8 Insinga RP, Itzler RF, Pellissier JM, Saddier P, Nikas AA: The incidence of herpes zoster in a United States administrative database. J Gen Intern Med 2005;20:748–753.

9 Veenstra J, Krol A, van Praag RM, Frissen PH, Schellekens PT, Lange JM, Coutinho RA, van der Meer JT: Herpes zoster, immunological deterioration and disease progression in HIV-1 infection. AIDS 1995;9:1153–1158.

10 Meigel W, Kern P, Dietrich M: Dermatologic diseases in patients with AIDS. Hautarzt 1986;37:250–254.

11 Rigopoulos D, Paparizos V, Katsambas A: Cutaneous markers of HIV infection. Clin Dermatol 2004;22:487–498.

12 Schöfer H: Frühsymptome der HIV-Erkrankung an Haut- und Schleimhäuten. AIDS Forsch 1991;6:633–648.

13 Garbe C, Husak R, Orfanos CE: HIV-associated dermatoses and their prevalence in 456 HIV-infected patients. Relation to immune status and its importance as a diagnostic marker. Hautarzt 1994;45:623–629.

14 Hartmann M, Petzoldt D: Viruserkrankungen der Haut bei HIV-Infizierten. Allergologie 1994;17:316–320.

15 Disler RS, Dover JS: Chronic localized herpes zoster in the acquired immunodeficiency syndrome. Arch Dermatol 1990;126:1105.

16 Schöfer H, Baur SI, Gregel C, Wolter M, Milbradt R: Hyperkoratotic varicella zoster virus infection in an HIV-infected patient. Successful treatment of persistent lesions with cryosurgery. Br J Dermatol 1998;138:714–715.

17 Onunu AN, Uhunwangho A: Clinical spectrum of herpes zoster in HIV-infected versus non-HIV infected patients in Benin City, Nigeria. West Afr J Med 2004;23:300–304.

18 Weber DM, Pelecchia JA: Varicella pneumonia: study of prevalence in adult men. JAMA 1965;192:572–577.

19 Alessi E, Cusini M, Zerboni R, Cavicchini S, Uberti-Foppa C, Galli M, Moroni M: Unusual varicella zoster virus infection in patients with the acquired immunodeficiency syndrome. Arch Dermatol 1988;124:1011–1013.

20 Gilson IH, Barnett JH, Conant MA, Laskin OL, Williams J, Jones PG: Disseminated ecthymatous herpes varicella-zoster virus infection in patients with acquired immunodeficiency syndrome. J Am Acad Dermatol 1989;20:637–642.

21 Hoppenjans WB, Bibler MR, Orme RL, Solinger AM: Prolonged cutaneous herpes zoster in acquired immunodeficiency syndrome. Arch Dermatol 1990;126:1048–1050.
22 Agusto V, Franca I, Mansinho K, Araujo C, Borges F, Champalimaud JL, Poiares-Baptista A, Martins C, Ricardo JL: Verrucous herpes zoster in AIDS patients. Acta Med Port 1997;10:497–501.
23 Jeyaratnam D, Robson AM, Hextall JM, Wong W, MacMahon E: Concurrent verrucous and varicelliform rashes following renal transplantation. Am J Transplant 2005;5:1777–1780.
24 Lewallen S: Herpes zoster ophthalmicus in Malawi. Ophthalmology 1994;101:1801–1804.
25 Sellitti TP, Huang AJW, Schiffman J, Davis JL: Association of herpes zoster ophthalmicus with acquired immunodeficiency syndrome. Am J Ophthalmol 1993;116:297–301.
26 Margolis TP, Milner MS, Shama A, Hodge W, Seiff S: Herpes zoster ophthalmicus in patients with human immunodeficiency virus infection. Am J Ophthalmol 1998;125:285–291.
27 Choo PW, Galil K, Donahue JG, Walker AM, Spiegelman D, Platt R: Risk factors for postherpetic neuralgia. Arch Intern Med 1997;157:1217–1224.
28 Kost RG, Straus SE: Postherpetic neuralgia – pathogenesis, treatment and prevention. N Engl J Med 1996;335:23–42.
29 Aldeen T, Hay P, Davidson F, Lau R: Herpers zoster infection in HIV-seropositive patients associated with highly active antiretroviral therapy. AIDS 1998;12:1719–1720.
30 Domingo P, Torres OH, Ris J, Vazquez G: Herpes zoster as an immune reconstitution disease after initiation of combination antiretroviral therapy in patients with human immunodeficiency virus type-1 infection. Am J Med 2001;110:605–609.
31 Dunic I, Djurkovic-Djakovic O, Vesic S, Zerjav S, Jevtovic D: Herpes zoster as an immune restoration disease in AIDS patients during therapy including protease inhibitors. Int J STD AIDS 2005;16:475–478.
32 Jevtovic DJ, Salemovic D, Ranin J, Pesic I, Zerjav S, Djurkovic-Djakovic O: The prevalence and risk of immune restoration disease in HIV-infected patients treated with highly active antiretroviral therapy. HIV Med 2005;6:140–143.
33 Martinez E, Gatell ME, Moran Y, Aznar E, Buira E, Guelar A, Mallolas J, Soriano E: High incidence of herpes zoster early after starting antiretroviral therapy with a protease inhibitor. Fifth Conference on Retroviruses and Opportunistic Infections. Chicago, February 1998, Abstract 501.
34 Alliegro MB, Dorrucci M, Pezzotti P, Rezza G, Sinicco A, Barbanera M, Castelli F, Tarantini G, Petrucci A: Herpes zoster and progression to AIDS in a cohort of individuals who seroconverted to human immunodeficiency virus. Italian HIV Seroconversion Study. Clin Infect Dis 1996;23: 990–995.
35 McNulty A, Li Y, Radtke U, Kaldor J, Rohrsheim R, Cooper DA, Donovan B: Herpes zoster and the stage and prognosis of HIV-1 infection. Genitourin Med 1997;73:467–470.
36 Moss AR, Bacchetti P, Osmond D, Krampf W, Chaisson RE, Stites D, Wilber J, Allain JP, Carlson J: Seropositivity for HIV and the development of AIDS or AIDS related condition: three year follow-up of the San Francisco General Hospital cohort. BMJ 1988;296:745–750.
37 Rogues AM, Dupon M, Ladner J, Ragnaud JM, Pellegrin JL, Dabis F: Herpes zoster and human immunodeficiency virus infection: a cohort study of 101 coinfected patients. Groupe d'Epidemiologie clinique du SIDA en Aquitaine. J Infect Dis 1993;45.
38 Galagher JG, Merigan TC: Prolonged herpes zoster infection associated with immunosuppressive therapy. Am Intern Med 1979;91:842–846.
39 Gourishankar S, McDermid JC, Jhangri GS, Preiksaitis JK: Herpes zoster infection following solid organ transplantation: incidence, risk factors and outcomes in the current immunosuppressive era. Am J Transplant 2004;4:108–115.
40 Lauzurica R, Bayes B, Frias C, Fontsere N, Hernandez A, Matas L, Jimenez A, Bonet J, Romero R: Disseminated varicella infection in adult renal allograft recipients: role of mycophenolate mofetil. Transplant Proc 2003;35:1758–1759.
41 Metzner B, Gruneisl R, Gebauer W, Reschke D, Ost E, Muller TH, Reichert D, Rosien B, Del Valle F, Zirpel I, Kohse KP, Schunter F, Illiger HJ: Late infectious complications after high-dose therapy and autologous blood stem cell transplantation. Med Klin 2002;97:650–658.
42 Rusthoven JJ, Ahlgren P, Elhakim T, Pinfold P, Reid J, Stewart L, Feld R: Varicella zoster infection in adult cancer patients: a population study. Arch Intern Med 1988;148:1561–1566.

43 Zambelli A, Montagna D, Da Prada GA, Maccario R, Zibera C, Moretta A, Ponchio L, Lozza L, Baiardi P, Maserati R, Marone P, Della Cuna GR: Evaluation of infectious complications and immune recovery following high-dose chemotherapy (HDC) and autologous peripheral blood progenitor cell transplantation (PBPC-T) in 148 breast cancer patients. Anticancer Res 2002;22: 3701–3708.
44 Dolin R, Reichman RC, Mazur MH, Whitley RJ. Herpes zoster-varicella infections in the immunosuppressed patients. Ann Intern Med 1978;89:375–388.
45 De Jong MD, Weel JF, van Oers MH, Boom R, Wertheim-van Dillen PM: Molecular diagnosis of visceral herpes zoster. Lancet 2001;357:2101–2102.
46 Itoh M, Kawaguchi S, Yago K, Shimada H, Muro H: Acute abdominal pain preceding cutaneous manifestations of varicella zoster infection after allogeneic bone marrow transplantation. Tohoku J Exp Med 2001;195:61–63.
47 O'Loughlin CJ, Karnam US, Barkin JS: Visceral varicella zoster after bone marrow transplantation: an obscure cause of an 'acute abdomen'. Dig Dis Sci 2002;47:1962–1964.
48 Guess HA, Broughton DD, Melton LJ 3rd, Kurland LT: Epidemiology of herpes zoster in children and adolescents: a population-based study. Pediatrics 1985;76:512–517.
49 Kanda Y, Mineishi S, Saito T, Saito A, Yamada S, Ohnishi M, Chizuka A, Niiya H, Suenaga K, Nakai K, Takeuchi T, Makimoto A, Tanosaki R, Kami M, Tanaka Y, Fujita S, Watanabe T, Kobayashi Y, Tobinai K, Takaue Y: Long-term low-dose acyclovir against varicella-zoster virus reactivation after allogeneic hematopoietic stem cell transplantation. Bone Marrow Transplant 2001;2:689–692.
50 Sullivan KM, Dykewicz CA, Longworth DL, Boeckh M, Baden LR, Rubin RH, Sepkowitz KA: Preventing opportunistic infections after hematopoietic stem cell transplantation: the Centers for Disease Control and Prevention, Infectious Diseases Society of America, and American Society for Blood and Marrow Transplantation Practice Guidelines and beyond. Hematology (Am Soc Hematol Educ Program) 2001;2001:392–421.
51 Ragozzino MW, Melton LJ 3rd, Kurland LT, Chu CP, Perry HO: Population-based study of herpes zoster and its sequelae. Medicine 1982;61:310–316.
52 Mazur MH, Dolin R. Herpes zoster at the NIH: a 20 year experience. Am J Med 1978;65: 738–744.
53 Gross G, Schöfer H, Wassilew S, Friese K, Timm A, Guthoff R, Pau HW, Malin JP, Wutzler P, Doerr HW: Herpes zoster guideline of the German Dermatology Society (DDG). J Clin Virol 2003;26:277–289.
54 Tyring S, Belanger R, Bezwoda W, Ljungman P, Boon R, Saltzman RL: A randomized, double-blind trial of famciclovir versus acyclovir for the treatment of localized dermatomal herpes zoster in immunocompromised patients. Cancer Invest 2001;19:13–22.
55 Breton G, Bouldouyre MA, Gervais A, Duval X, Longuet P, Leport C, Vilde JL: Failure of valacyclovir for herpes zoster in a moderately immunocompromised HIV-infected patient. AIDS Patient Care STDS 2004;18:255–257.

Helmut Schöfer, MD
Department of Dermatology, University Hospital, J.W. Goethe University
Theodor Stern Kai 7
DE–60590 Frankfurt (Germany)
E-Mail schoefer@em.uni-frankfurt.de

Gross G, Doerr HW (eds): Herpes Zoster.
Monogr Virol. Basel, Karger, 2006, vol 26, pp 107–116

Chickenpox and Zoster in Marrow Transplant Recipients

I. Genvresse[a], *G. Maschmeyer*[b]

[a]Department of Haematology and Oncology, Charité University Hospital, Campus Virchow-Klinikum, Berlin; [b]Department of Haematology and Oncology, Klinikum Ernst von Bergmann, Potsdam, Germany

Incidence, Onset and Risk Factors of Varicella–Zoster Virus Infection in Marrow Transplant Recipients

Varicella–zoster virus (VZV) is one of the most common infectious agents encountered in patients undergoing bone marrow transplantation (BMT), mostly through reactivation of latent virus. Both allogeneic and autologous BMT patients have an increased incidence of VZV infections. In two large retrospective series, VZV infection among allogeneic BMT patients occurred with an incidence of 21 and 41%, respectively [1, 2]. In autologous BMT recipients, due to the less degree of the immunological impairment experienced by these patients, the incidence of VZV infection is generally lower than among allogeneic recipients, averaging 15–25% [3], although some authors reported higher incidence rates [4]. Most VZV infections occur within the first year after transplantation. Typically, VZV reactivation develops after a median of 5–7 months post-BMT. In the study by Koc et al. [2], 68% of VZV infections emerged within 1 year after transplantation, however, only 12% were observed within the first 3 months. In the study by Steer et al. [1], who used a prolonged antiviral prophylaxis with low-dose acyclovir (ACV) and/or ganciclovir (GCV) given for at least 6 months after allogeneic transplantation, a delay in the onset of VZV reactivation to a median onset of 16 months post-BMT was reported (table 1).

Various putative risk factors have been studied to identify patient groups at increased risk of VZV reactivation. Of these, age ≥10 years [4, 5], VZV seropositivity [4], pre-transplant total body or total lymphatic irradiation [4], presence of acute or extensive chronic GVHD [1, 2, 5] and the use of post-transplant antithymocyte globulin [5] have been associated with a higher incidence of

Table 1. Incidence of varicella–zoster virus (VZV) infections after bone marrow or stem cell transplantation

Year	Reference	Underlying disease	Transplant type	Patients n	VZV infection (%)
1980		Leukaemia	Allogeneic/syngeneic	33	21
1982		Leukaemia/aplastic anaemia	Allogeneic/syngeneic	98	52
1985	[5]	Leukaemia/aplastic anaemia	Allogeneic	1,394	17
1986		Haematol. malignancy	Allogeneic	73	36
1989		Leukaemia/solid tumours	Allogeneic	236	23
1989	[6]	Leukaemia/lymphoma	Autologous	153	28
1991		Hodgkin's disease	Autologous	28	32
1992		Leukaemia/lymphoma/other	Autologous/allogeneic	51	31
1999	[15]	Leukaemia/lymphoma/other	Autologous	215	19
2000	[2]	Leukaemia/lymphoma/other	Allogeneic	100	41
2000	[1]	Leukaemia/aplastic anaemia	Allogeneic	151	21
2001	[3]	Leukaemia/lymphoma/other	Autologous	164	16

Comprised according to [13].

VZV reactivation. Underlying lymphoproliferative disorders may be a risk factor for BMT recipients in contrast to patients with solid tumours or chronic myeloid leukaemia [5, 6], however, a 23% incidence of VZV infection was found among women with breast cancer undergoing intensive chemotherapy and autologous stem cell transplantation (SCT) [7]. The association of primary VZV infection and the pretransplant VZV serology was analysed in a retrospective study of VZV infection after SCT in 109 consecutive children, 96 of whom receiving allogeneic transplants. Pretransplant VZV seropositivity in recipients was the only significant risk factor for the development of Zoster within 1 year after transplantation in a multivariate analysis [8]. In this study, 18% of the posttransplant VZV infections were chickenpox [8].

Reconstitution of VZV-Specific T-Cell Immunity after Haematopoietic Stem Cell Transplantation

The escape of VZV from immune surveillance is associated with declining numbers of VZV-specific memory T cells in the latently infected host. Consequently, among SCT recipients, as well as among other patients with impaired cellular immunity such as AIDS patients, prolonged T cell depletion and the resulting reduction or loss of VZV-specific T cells, demonstrated by assays for cytotoxic function as well as by in vitro T lymphocyte proliferation to VZV antigens, predict a high risk for VZV reactivation [9–11]. In a study of immune reconstitution in patients receiving allogeneic bone marrow or G-CSF-mobilised peripheral blood SCT, the incidence of viral infections was not significantly different. Apart from that, in vitro VZV-induced proliferation assays suggest that the $CD4^+$ T cell function was similar in the two groups [12]. So far, there are no data on the reconstitution of VZV-specific T-cell immunity following allogeneic SCT in recipients conditioned by reduced-intensity vs. myeloablative regimens.

Autologous SCT recipients can be separated into distinct risk groups on the basis of their $CD4^+$ and $CD8^+$ lymphocyte counts: patients with $CD4^+$ cells $<200/\mu l$ and $CD8^+$ cells $<800/\mu l$ at day 30 post-transplant had a risk of VZV reactivation of 48% at 1 year [3].

Reconstitution of VZV memory T cells can usually be detected at 9–12 months after SCT and corresponds to a reduction in the risk of Zoster and its complications [9, 13]. VZV reactivation, resulting in in vivo re-exposure to viral antigens, plays a major role in inducing this immune reconstitution. Meyers et al. [9] detected T-lymphocyte recognition of VZV antigens by proliferation assay in 16 of 18 BMT patients (89%) after symptomatic recurrences of VZV vs. 15 of 29 patients (51%) who did not develop zoster. In SCT recipients with recovered VZV immunity, yet without clinical signs of VZV recurrence, episodes of VZV viraemia may provide the stimulus to restore adaptive immunity to VZV [11]. Thus, suppression of VZV by antiviral prophylaxis may delay the reactivation and subsequently VZV-specific immune reconstitution [1, 10].

The observation of long-lasting VZV-specific cellular immunity in patients who experienced VZV infection in the immediate pre-transplant period [14], along with the evidence that natural VZV reactivation induces recovery of VZV-specific T cells constitute the basis to develop vaccination strategies in SCT recipients.

Clinical Characteristics of VZV Infection in Marrow Transplant Recipients

Localised Herpes Zoster

Typical vesicles on an erythematous background with a dermatomal distribution are the most common clinical presentation of VZV infection in SCT recipients, accounting for up to 90% of cases [1–6, 15]. Localised Zoster is defined as fewer than 6 vesicles beyond the boundaries of the primary dermatome. Rash is usually preceded by radicular pain and paraesthesia or dysaesthesia in the involved dermatome, so-called pre-herpetic neuralgia. The most frequently involved sites include thoracic and lumbosacral nerve dermatomes followed by cranial and cervical dermatomes [2, 3, 8, 15]. Delayed or inadequate host response results in an average time for cessation of new lesion formation of 8 days, compared with 3–5 days in the immunocompetent host. The time to complete healing is also prolonged.

Local scarring and bacterial superinfection are also more common in SCT recipients [2, 5, 6, 15] and significantly more frequent among patients with VZV during the first 9 months after transplant than among patients with late infections [5]. However, postherpetic neuralgia remains the most common complication, observed in 32–41% of affected SCT recipients [2, 3, 14], who therefore appear to be at higher risk compared to immuncompetent individuals [5].

Some SCT patients have a chronic cutaneous VZV reactivation, indicating continued viral replication in ganglia and skin and a failure of the host response to overcome the immune evasion mechanisms of the virus.

Cutaneous and Visceral Dissemination

Disseminated Zoster consists of 6 skin lesions beyond the boundaries of the primary dermatome and/or visceral involvement. Before the introduction of VZV prophylaxis, 45% of SCT recipients experienced cutaneous and visceral dissemination with an overall mortality of up to 10% [5]. More recent studies using antiviral prophylaxis showed incidences of cutaneous dissemination of 3.8–25% among autologous SCT recipients [3, 15] and 17–38% after allogeneic BMT [1, 2], but a very low rate of visceral dissemination (0–7%). Consequently, the mortality from VZV infection after BMT is now below 1%.

The most common sites of VZV organ infection are lungs and liver, resulting in pneumonia, hepatitis and intravascular coagulopathy, but myelitis and meningoencephalitis have been reported as well [16]. The fatality of these disseminated visceral forms approaches 30–50%. Fatal disseminated VZV infections are most often due to pneumonia [4].

There has been a small number of case reports of BMT patients where visceral disseminated disease was not preceded by any skin involvement [17, 18], demonstrating that viraemia can occur without replication of the virus in the skin, presumably by entry of virus into T cells that traffic through sensory ganglia. These rare presentations may remain undiagnosed.

Second Episodes of Zoster

Some SCT recipients have recurrent zoster within days after discontinuation of antiviral therapy, indicating a failure to re-establish latency in the short-term.

Chickenpox

Chickenpox is defined as generalised onset of vesicular skin eruptions without dermatomal localisation in a VZV seronegative patient without clinical history of previous chickenpox. Before the introduction of ACV prophylaxis, chickenpox emerged in 25% of children during the first 6 months after BMT [19]. After introduction of low-dose ACV and/or GCV prophylaxis, this rate has dropped to 5.5% [8].

However, because the absence of previous VZV infection (history of chickenpox and presence of VZV-specific antibodies) does not exclude the possibility of post-transplant VZV reactivation [8], chickenpox cannot be distinguished from disseminated cutaneous Zoster, so that both clinical entities are often referred as 'varicella-like' cutaneous involvement accounting for 17% in adult SCT recipients [11].

Management of VZV Infections in Marrow Transplant Recipients

Diagnosis

The diagnosis of VZV infection is established by clinical findings. However, confirmation by rapid, sensitive laboratory methods such as immunofluorescence, enzyme immunoassay or PCR is essential in many patients due to unusual clinical manifestations of VZV infection in this population. Serologic diagnosis of VZV reactivation is not reliable in SCT patients because the rise of VZV IgG antibodies or detection of IgM antibodies is delayed or absent. Detection of a PCR product after amplification of VZV DNA from peripheral blood mononuclear cells or from plasma or serum to document VZV viraemia in SCT patients appears compelling. In a retrospective survey, transient VZV viraemia was documented with this method in 19% of the allogeneic BMT recipients [2]. In cases of visceral disseminated VZV infection without skin involvement, VZV DNA can be detected in the serum by PCR [18, 20].

Antiviral Treatment

The recommended therapy for chickenpox or zoster in BMT patients is intravenous (iv) ACV at a dosage of 10 mg/kg or 500 mg/m^2 every 8 h for at least 7 days. This treatment has been shown to prevent the dissemination of the disease, even when it is initiated only 6 days after the onset of rash [21, 22]. For localised dermatomal zoster, oral ACV, 800 mg every 4–5 h, was compared with i.v. ACV in a small randomised study in allogeneic SCT patients, and the outcome was comparable [23]. Oral famciclovir in a dosage of 500 mg every 8 h was compared with oral ACV (800 mg q 4–5 h) in SCT, solid organ transplant and oncological patients, and the results indicated similar efficacy [24]. No controlled study has been performed with valacyclovir given for the treatment of zoster in SCT recipients, but it is likely that valacyclovir would have comparable antiviral effects in this patient group as well.

VZV resistance to ACV is rare but has been reported after BMT [25]. In some cases, ACV-resistant VZV strains emerge during therapy, particularly under prolonged, low-dose regimens. Because most resistant VZV strains show mutations in thymidine kinase, ACV, valacyclovir, and famciclovir will be ineffective.

VZV-Immunoglobulin

Varicella–zoster immune globulin is the recommended prophylactic measure in seronegative SCT recipients after a close or household exposure to varicella, if it can be given within 4 days of exposure [26]. Varicella–zoster immune globulin is not used in patients with zoster, because VZV IgG titres are maintained despite the loss of cell-mediated immunity to the virus. There is no evidence that antibody prophylaxis increases antibody titres or reduces the risk of VZV reactivation after SCT.

Antiviral Prophylaxis

Because the median onset of VZV reactivation is 5–7 months after SCT, the duration of antiviral prophylaxis must be long enough to prevent VZV disease. A summary of studies of long-term ACV prophylaxis after BMT is presented in table 2. Two randomised, controlled studies have been performed comparing 6 months of prophylactic ACV with placebo: in the first one, i.v. ACV (250 mg/m^2 every 8 h, starting 5 days before transplantation) for 5 weeks and followed by oral ACV (1,200 mg daily) prevented zoster in allogeneic BMT patients [27]. Similar results were obtained in a study on higher doses of ACV (3,200 mg daily) [28]. Both studies, along with a trial in autologous BMT recipients [29], showed that although reactivation was suppressed during the period of prophylaxis, no overall reduction of the VZV reactivation rate could be achieved 12 months after transplantation, due to delayed onset after ACV discontinuation. Possibly, prolonged (>6 months) oral antiviral prophylaxis after transplantation could reduce the incidence of VZV infection post-BMT, allowing patients to restore their cellular immunity. This was studied by Kanda et al. [30], using low-dose ACV (400 mg p.o. daily) continued until the end of immunosuppressive therapy after SCT. Thereby, the first-year cumulative incidence of VZV reactivation could be reduced to 29%. Low-dose prophylaxis might permit subclinical VZV reactivation, leading to an immune recovery against VZV. The major concern against prolonged low-dose ACV prophylaxis is the emergence of VZV strains resistant to ACV. However, in studies on low-dose ACV prophylaxis, no breakthrough reactivation of VZV could be observed and symptomatic VZV reactivation after discontinuing ACV was always successfully treated with a therapeutic dose of ACV [1, 30]. Another concern is the cost of long-term prophylaxis.

Table 2. Results of published trials on long-term acyclovir prophylaxis of VZV reactivation after SCT

Reference	Transplant type	Daily dose	Duration	Reactivation after discontinuation of prophylaxis
Ljungman [27]	Allogeneic	1,200 mg p.o.	6 months	31% at 6 months
Selby et al. [28]	Allogeneic	3,200 mg p.o.	6 months	45% at 6 months
Sempere et al. [29]	Autologous	1,200 mg p.o.	6 months	24% at 6 months
Steer et al. [1]	Allogeneic	600 mg p.o.	6 months	33% at 12 months
Kanda et al. [30]	Allogeneic	400 mg p.o.	During immuno-suppression	29% at 12 months

From reference [30].

Due to its toxic potential, GCV has not been evaluated for prophylaxis of VZV infection. However, patients receiving GCV as CMV prophylaxis had delayed onset of VZV reactivation [1, 2]. Valacyclovir has not been studied for VZV prophylaxis, but the rate of VZV disease was reduced in a study when valacyclovir was compared to ACV as CMV prophylaxis [31]. There is no published data on VZV prophylaxis with famciclovir.

Vaccination

One of the new strategies to control shingles effectively is immunotherapy. Recent studies have obtained encouraging results. Redman et al. [32] examined the impact of the heat-inactivated varicella vaccine in 75 patients who had undergone allogeneic or autologous SCT. At 4 months post-transplant, patients vaccinated 1–3 months after BMT demonstrated a 4-fold increase in their stimulation index compared with unvaccinated patients. The incidence of VZV reactivation did not differ between the 2 groups, but disease severity was significantly reduced in the vaccinated cohort. In lymphoma patients undergoing SCT, a fourth dose was given within 30 days before transplantation, and the incidence of Zoster was significantly reduced in vaccinated patients compared to unvaccinated patients (13 vs. 33%) [33].

Although the available live attenuated varicella vaccine is not approved for administration post-SCT, Sauerbrei et al. [34] studied 15 children immunised against varicella between 12 and 23 months post-transplant. Eight of 9 seronegative patients seroconverted, and 6 patients maintained their virus-specific IgG 2 years after vaccination.

A live attenuated Oka/Merck VZV vaccine ('zoster vaccine') has now been successfully tested for prevention of varicella and reduction of the burden-of-illness caused by herpes zoster in a large placebo-controlled trial in healthy adults [35]. To what extent the inactivated vaccines will protect patients after stem cell transplantation from zoster has to be addressed.

Recovery of VZV immunity in BMT patients might be further improved by immunising donors and recipients before allogeneic transplantation or by initiating the vaccine regimen at a shorter interval after transplantation.

References

1 Steer CB, Szer J, Sasadeusz J, Matthews JP, Beresford JA, Grigg A: Varicella-zoster infection after allogeneic bone marrow transplantation: incidence, risk factors and prevention with low-dose aciclovir and ganciclovir. Bone Marrow Transplant 2000;25:657–664.

2 Koc Y, Miller KB, Schenkein DP, Griffith J, Akhtar M, DesJardin J, Snydman DR: Varicella zoster virus infections following allogeneic bone marrow transplantation: frequency, risk factors, and clinical outcome. Biol Blood Marrow Transplant 2000;6:44–49.

3 Offidani M, Corvatta L, Olivieri A, Mele A, Brunori M, Montanari M, Rupoli S, Scalari P, Leoni P: A Predictive model of varicella-zoster virus infection after autologous peripheral blood progenitor cell transplantation. Clin Infect Dis 2001;32:1414–1422.

4 Han CS, Miller W, Haake R, Weisdorf D: Varicella zoster infection after bone marrow transplantation: incidence, risk factors and complications. Bone Marrow Transplant 1994;4:191–194.

5 Locksley RM, Flournoy N, Sullivan KM, Meyers JD: Infection with varicella-zoster virus after marrow transplantation. J Infect Dis 1985;152:1172–1181.

6 Schuchter LM, Wingard JR, Piantadosi S, Burns WH, Santos GW, Saral R: Herpes zoster infection after autologous bone marrow transplantation. Blood 1989;74:1424–1427.

7 Chandrasekar PH, Abraham OC, Klein J, Alangaden G, Chalasani G, Cassells L, Dansey R, Abella S, Karanes C, Peters W, Baynes R: Low infectious morbidity after intensive chemotherapy and autologous peripheral blood progenitor cell transplantation in the outpatient setting for women with breast cancer. Clin Infect Dis 2001;32:546–551.

8 Leung TF, Chik KW, Li CK, Lai H, Shing MMK, Chan PKS, Lee V, Yuen PMP: Incidence, risk factors and outcome of varicella-zoster virus infection in children after haematopoietic stem cell transplantation. Bone Marrow Transplant 2000;25:167–172.

9 Meyers JD, Flournoy N, Thomas ED: Cell-mediated immunity to varicella-zoster virus after allogeneic marrow transplant. J Infect Dis 1980;141:479–487.

10 Ljungman P, Lonnqvist B, Gahrton G, Ringden O, Sundqvist VA, Wahren B: Clinical and subclinical reactivations of varicella-zoster virus in immunocompromised patients. J Infect Dis 1986;153: 840–847.

11 Wilson A, Sharp M, Koropchak CM, Ting SF, Arvin AM: Subclinical varicella-zoster virus viremia, herpes zoster, and T lymphocyte immunity to varicella-zoster viral antigens after bone marrow transplantation. J Infect Dis 1992;165:119–126.

12 Storek J, Dawson MA, Storer B, Stevens-Ayers T, Maloney DG, Marr KA, Witherspoon RP, Bensinger W, Flowers MED, Martin P, Storb R, Appelbaum FR, Boeckh M: Immune reconstitution after allogeneic marrow transplantation compared with blood stem cell transplantation. Blood 2001;97:3380–3389.

13 Arvin AM: Varicella-zoster virus: pathogenesis, immunity, and clinical management in hematopoietic cell transplant recipients. Biol Blood Marrow Transplant 2000;6:219–230.

14 Kato S, Yabe H, Yabe M, Kimura M, Ito M, Tsuchida F, Tsuji K, Takahashi M: Studies on transfer of varicella-zoster virus specific T-cell immunity from bone marrow donor to recipient. Blood 1990;75:806–809.

15 Bilgrami S, Chakraborty NG, Rodriguez-Pinero F, Khan AM, Feingold JM, Bona RD, Edwards RL, Dorsky D, Clive J, Mukherji B, Tutschka PJ: Varicella zoster virus infection associated with high-dose chemotherapy and autologous stem-cell rescue. Bone Marrow Transplant 1999;23: 469–474.

16 Tauro S, Toh V, Osman H, Mahendra P: Varicella zoster meningoencephalitis following treatment for dermatomal zoster in an alloBMT patient. Bone Marrow Transplant 2000;26:795–796.

17 Stemmer SM, Kinsman K, Tellschow S, Jones RB: Fatal noncutaneous visceral infection with varicella-zoster virus in a patient with lymphoma after autologous bone marrow transplantation. Clin Infect Dis 1993;16:497–499.

18 Yagi T, Karasuno T, Hasegawa T, Yasumi M, Kawamoto S, Murakami M, Uosima N, Nakamura H, Hiraoka A, Masaoka T: Acute abdomen without cutaneous signs of varicella zoster virus infection as a late complication of allogeneic bone marrow transplantation: importance of empiric therapy with acyclovir. Bone Marrow Transplant 2000;25:1003–1005.

19 Wacker P, Hartmann O, Benhamou E, Salloum E, Lemerle J: Varicella-zoster virus infections after autologous bone marrow transplantation in children. Bone Marrow Transplant 1989;4:191–194.

20 Rogers SY, Irving W, Harris A, Russel NH: Visceral varicella zoster infection after bone marrow transplantation without skin involvement and the use of PCR for diagnosis. Bone Marrow Transplant 1995;15:805–807.

21 Balfour HH Jr, Bean B, Laskin OL, Ambinder RF, Meyers JD, Wade JC, Zaia JA, Aeppli D, Kirk LE, Segreti AC, Keeney RE: Acyclovir halts progression of herpes zoster in immunocompromised patients. N Engl J Med 1983;308:1448–1453.

22 Shepp DH, Dandliker PS, Meyers JD: Treatment of varicella-zoster virus infection in severely immunocompromised patients. A randomised comparison of acyclovir and vidarabine. N Engl J Med 1986;314:208–212.

23 Ljungman P, Lönnqvist B, Ringdén O, Skinhöj P, Gahrton G: A randomised trial of oral versus intravenous acyclovir for treatment of herpes zoster in bone marrow transplant recipients. Nordic Bone Marrow Transplant Group. Bone Marrow Transplant 1989;4:613–615.

24 Tyring S, Belanger R, Bezwoda W, Ljungman P, Boon R, Saltzman RL: A randomised, double-blind trial of famciclovir versus acyclovir for the treatment of localized dermatomal herpes zoster in immunocompromised patients. Cancer Invest 2001;19:13–22.

25 Reusser P, Cordonnier C, Einsele H, Engelhard D, Link H, Locasciulli A, Ljungman P, for the Infectious Diseases Working Party of the European Group for Blood and Marrow Transplantation (EBMT): European survey of herpesvirus resistance to antiviral drugs in bone marrow transplant recipients. Bone Marrow Transplant 1996;17:813–817.

26 Centers for Disease Control (CDC): Guidelines for preventing opportunistic infections among hematopoietic stem cell transplant recipients. Recommendations of the CDC, the Infectious Disease Society of America, and the American Society of Blood and Marrow Transplantation. MMWR Morb Mortal Wkly Rep 2000;49:1–125.

27 Lundgren G, Wilczek H, Lönnqvist B, Lindholm A, Wahren B, Ringdén O: Acyclovir prophylaxis in bone marrow transplant recipients. Scand J Infect Dis 1985;47(suppl):137–144.

28 Selby PJ, Powles RL, Easton D, Perren TJ, Stolle K, Jameson B, Fiddian AP, Tryhorn Y, Stern H: The prophylactic role of intravenous and long-term oral acyclovir after allogeneic bone marrow transplantation. Br J Cancer 1989;59:434–438.

29 Sempere A, Sanz GF, Senent L, de la Rubia J, Jarque I, López F, Arilla MJ, Guinot M, Martín G, Martínez K, Marty ML, Sanz MA: Long-term acyclovir for prevention of varicella zoster virus infection after autologous blood stem cell transplantation in patients with acute leukemia. Bone Marrow Transplant 1992;10:495–498.

30 Kanda Y, Mineishi S, Saito T, Saito A, Yamada S, Ohnishi M, Chizuka A, Niiya H, Suenaga K, Nakai K, Takeuchi T, Makimoto A, Tanosaki R, Kami M, Tanaka Y, Fujita S, Watanabe T, Kobayashi Y, Tobinai K, Takaue Y: Long-term acyclovir against varicella-zoster virus reactivation after allogeneic hematopoietic stem cell transplantation. Bone Marrow Transplant 2001;28: 689–692.

31 Ljungman P, Aschan J, Lewensohn-Fuch I, Carlens S, Larsson K, Lönnqvist B, Mattsson J, Sparrelid E, Winiarski J, Ringdén O: Results of different strategies for reducing cytomegalovirus-associated mortality in allogeneic stem cell transplant recipients. Transplantation 1998;66: 1330–1334.

32 Redman RL, Nader S, Zerboni L, Liu C, Wong RM, Brown BW, Arvin AM: Early reconstitution of immunity and decreased severity of herpes zoster in bone marrow transplant recipients immunized with inactivated varicella vaccine. J Infect Dis 1997;176:578–585.

33 Hata A, Asanuma H, Rinki M, Sharp M, Wong RM, Blume K, Arvin AM: Use of an inactivated varicella vaccine in recipients of hematopoietic-cell transplants. N Engl J Med 2002;347:26–34.
34 Sauerbrei A, Prager J, Hengst U, Zintl F, Wutzler P: Varicella vaccination in children after bone marrow transplantation. Bone Marrow Transplant 1997;20:381–383.
35 Oxman MN, Levin MJ, Johnson GR, Shamder KE et al. A vaccine to prevent herpes zoster and postherpetic neuralgia in older adults. N Engl J Med 2005;352:2271–2284.

Georg Maschmeyer, MD
Professor of Internal Medicine, Director
Department of Haematology and Oncology, Klinikum Ernst von Bergmann
Charlottenstrasse 72
DE–14467 Potsdam (Germany)
Tel. +49 331 241 6001, Fax +49 331 241 6000, E-Mail gmaschmeyer@klinikumevb.de

Gross G, Doerr HW (eds): Herpes Zoster.
Monogr Virol. Basel, Karger, 2006, vol 26, pp 117–122

General Aspects of Therapy

G. Gross

Department of Dermatology and Venereology, University of Rostock,
Rostock, Germany

Patients suffering from herpes zoster should be encouraged to see a physician as early as possible for immediate medical care based on administration of systemic antiviral therapy. In addition symptomatic local therapy and analgetic therapy in order to achieve painlessness are equally important. Since years it has become clear, that systemic antiviral therapy is indicated for most patients suffering from herpes zoster. In general the aims of therapy for herpes zoster comprise the following: decrease viral replication as early as possible, thus lowering the viral load, accelerate healing, limit or relieve severity and duration of acute and chronic pain (postherpetic neuralgia, PHN). Further options are to prevent or alleviate other acute and chronic herpes zoster complications and reduce the risk of cutaneous extension and visceral dissemination of VZV, which is particularly a problem in immunocompromised patients. Alternative therapies such as hypnosis and others are definitely of little help in terms of prevention of acute and chronic complications. Therefore, these therapies should be omitted. Patients with herpes zoster involving cranial nerves, such as ophthalmic zoster and zoster oticus, must be referred as early as possible for specialist ophthalmic and otorhinolaryngologic management. There is available evidence from clinical trials that early treatment with oral antiviral agents achieves many of the mentioned aims [1–3]. The risk of ocular complications is reduced from about 50% to 20–30% with oral antiviral therapy.

Antiviral Therapy

Systemic antiviral therapy is urgently indicated in herpes zoster patients beyond the age of 50 years and in patients of any age, including children, with herpes zoster affecting the head and neck area, especially zoster ophthalmicus

Table 1. Indications for systemic antiviral therapy of herpes zoster

Urgent indications
Herpes zoster at any site of the body in patients beyond the age of 50
Herpes zoster in the head/neck area of patients at any age
Severe herpes zoster on the trunk and on the limb
Herpes zoster in immunodeficient patients
Herpes zoster in patients with severe atopic dermatitis and severe eczema
Relative indications
Herpes zoster on the trunk/on the extremities in patients younger than 50 years

and zoster oticus (table 1). Severe herpes zoster of the trunk and extremities as well as herpes zoster in immunodeficient patients and patients with atopic dermatitis or other eczema are also indications for antiviral therapy. Furthermore, zoster ophthalmicus and zoster oticus as well as age over 50 years are major risk factors for the development of PHN. Some authorities consider antiviral treatment to be only optional in patients younger than 50 years of age with shingles on the trunk or on the extremities. Nevertheless, antiviral therapy has potential benefit and bears a limited risk. The same is true for immunocompetent children. However, as a rule patients younger than 50 years should be checked serologically for HIV-antibodies, because herpes zoster is an important marker for HIV-infection and AIDS.

In 1998, the Zoster Study Group of the Dermatological Research Cooperative Group (ADF) associated with the German Dermatology Society (DDG) published a PHN prevention zoster score, which has shown to be an useful aid for initiating systemic antiviral therapy [4, 5]. The zoster score is taking into account the following risk factors: age beyond 50 years, female gender, presence of more than 50 herpes zoster vesicles, haemorrhagic lesions and herpes zoster involvement of cranial and sacral dermatomes.

Four different antivirals (acyclovir, valacyclovir, famciclovir and brivudin) (fig. 1) have become available for the treatment of herpes zoster over the years. All of these are nucleoside analogues, which interfere with viral nucleic acid chain synthesis, thereby inhibiting viral replication. In more recent years, oral acyclovir has been replaced by its prodrug valacyclovir in the oral treatment of herpes zoster. Acyclovir has a comparable lower oral bioavailability (20%) than valacyclovir. Another nucleoside analogue is penciclovir, which has a similar activity and mechanism compared to acyclovir. Its prodrug famciclovir is used for oral treatment of VZV infections (herpes zoster and chickenpox). While

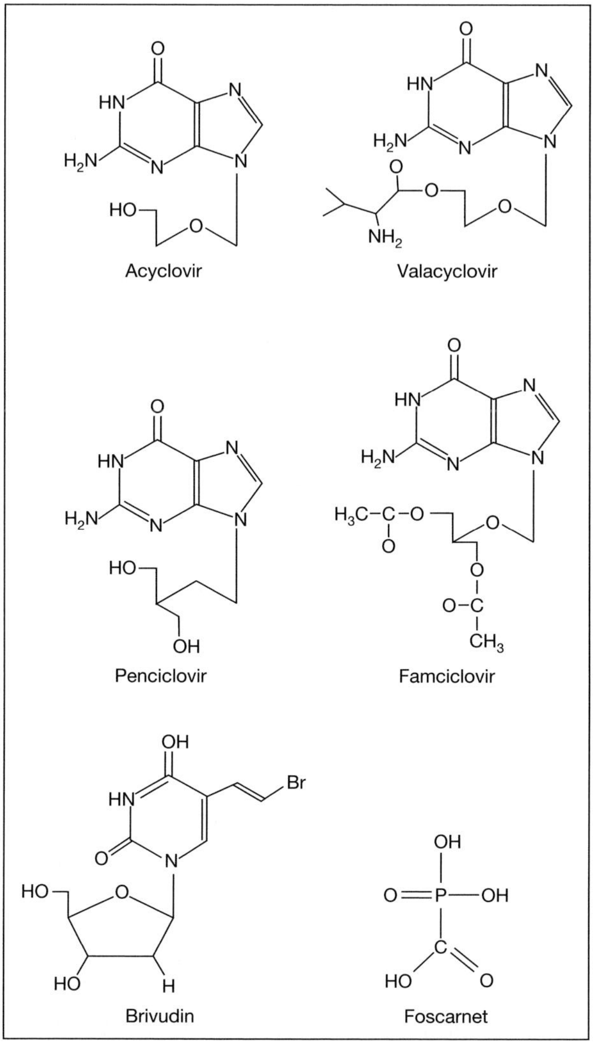

Fig. 1. Antivirals used in the treatment of herpes zoster.

acyclovir can be administered orally and parenterally, its prodrug valacyclovir, famcyclovir (prodrug of pencyclovir) as well as brivudin are exclusively administered orally. Brivudin, the most recently introduced antiviral compound for herpes zoster has a markedly higher anti-VZV potency than oral acyclovir, valacyclovir and famcyclovir (see chapter: Antiviral Therapy of Shingles in Dermatology, pp 123–130). All 4 drugs, however are well tolerated and do not differ widely in terms of efficacy and safety. However, because of improved pharmaco-kinetic properties and simpler dosing regimes, recently oral valacyclovir, famcyclovir and brivudin have been used preferentially in comparison to acyclovir. While acyclovir, valacyclovir and famcyclovir represent the common antiviral treatment used for herpes zoster so far, brivudin has become actually a frequently used antiviral compound for this indication particularly in European countries [6].

In order to achieve successful results in patients with herpes zoster, antiviral therapy should be started as early as possible (i.e. within 48 h to a maximum of 72 h after onset of rash) and should be continued strictly for 7 days. A major limitation of antiviral therapy is the delay between onset of symptoms and initiation of antiviral therapy. Although it is well accepted that the likelihood of beneficial effects is higher with earlier intervention [7], some patients will benefit from antiviral therapy even if it is started after 72 h. Patients with disseminated herpes zoster and evidence of immunodeficiency should be always treated antivirally even later than 72 h after onset of the rash.

When Intravenous Antiviral Therapy has to be Considered?

The standard antiviral treatment for herpes zoster in immunodeficient patients is acyclovir intravenously in a dose of 8–10 mg per kg body weight, 3 times daily during 7–14 days [8]. While intravenous acyclovir was compared with brivudin in a study of Wutzler et al. [9], no controlled studies comparing intravenous acyclovir with oral acyclovir, valacyclovir or famcyclovir have been performed in immunodeficient persons suffering from herpes zoster.

Other conditions where intravenous acyclovir therapy has to be considered are particularly herpes zoster ophthalmicus and herpes zoster oticus at any age. This applies also for children with herpes zoster affecting the VIth, the VIIth and the VIIIth cranial nerve. In view of the unpredictable course, herpes zoster in patients with HIV infection, herpes zoster in immunodeficient children and also in atopic children (prevention of scar formation) should be treated with acyclovir intravenously. Elderly and debilitated patients under multidrug therapy are a further group of persons that should be considered to be treated with acyclovir parenterally under stationary conditions. Herewith, harmful undertreatment due to mistakes taking antiviral drugs continuously is intended to prevent.

Symptomatic Therapy

Adequate control of acute symptoms is important for optimal therapy. So concurrent symptomatic treatment of cutaneous zoster lesions is always neccessary. Helpful measures for the local management include keeping lesions clean, dry and protected with a sterile nonadherent dressing. However, topical antiviral agents have shown to be ineffective and thus definitely have no place in the management of shingles [2, 3]. Topical lotions such as lotio alba, vioform zinc mixture etc. or crust removal of older lesions using an antisepic cream may help to control burning and itching which are common symptoms associated with acute stage herpes zoster.

Painlessness is the most important goal of therapy in the acute phase of herpes zoster. Consequent control of acute pain may prevent mechanisms thought to be related to the development of PHN [10]. It is anticipated that the risk of PHN might be reduced by a combined treatment consisting of a systemic antiviral drug and an appropriately-dosed analgesic, a tricyclic antidepressant, an antiepileptic or an adjunctive therapy such as gabapentin or pregabalin [2, 3]. So far, however it has not been demonstrated clearly, that systematic analgesia in acute herpes zoster prevents the development of PHN. Acute pain management in children with zoster can usually be accomplished with ibuprofen. For acute severe pain codeine may be helpful [11].

The Role of Corticisteroids in the Management of Herpes Zoster

Oral corticosteroids are not recommended for the treatment of herpes zoster [2, 3]. Some countries, however have approved them for combined therapy with oral acyclovir therapy [12] in patients not being at risk for corticosteroid-related complications.

Two large prospective studies [3, 13] showed that herpes zoster patients treated with a combination therapy consisting of oral acyclovir and prednisone had an improved quality of life and some improvement in acute herpes zoster symptoms, particularly in alleviation of acute pain. However, combination of oral acyclovir and corticosteroids had no effect on the incidence of postherpetic neuralgia. The available study results have been achieved with a combination therapy of oral acyclovir and oral prednisolone (60 mg per day, dose reduction over about 10–14 days, to 5 mg per day). As a conclusion, use of oral corticosteroids with systemic antiviral therapy, if considered must be very carefully outweighed with the risk of adverse effects [12].

References

1 Bansal R, Tutrone WD, Weinberg JM: Viral skin infections in the elderly: diagnosis and management. Drugs Aging 2002;19:503–514.

2 Gnann JW Jr, Whitley RJ: Clinical practice. Herpes zoster. N Engl J Med 2002;347:340–346.

3 Whitley RJ, Weiss H, Gnann W, Tyring S, Mertz GJ, Pappas PG, Schleupner CJ, Hayden F, Wolf J, Soong SJ: Acyclovir with and without prednisone for the treatment of herpes zoster. A randomized placebo-controlled trial. The National Institute of Allergy and Infectious Diseases. Collaborative Antiviral Study Group. Ann Intern Med 1996;125:376–383.

4 Meister W, Neiss A, Gross G, Doerr H, Holbel W, Malin J, von Essen J, Reimann B, Witke C, Wutzler P: Demography, symptomatology, and course of disease in ambulatory zoster patients. A physician-based survey in Germany. Intervirology 1998;41:272–277.

5 Meister W, Neiss A, Gross G, Doerr HW, Holbel W, Malin JP, von Essen J, Reimann BY, Witke C, Wutzler P: A prognostic score for postherpetic neuralgia in ambulatory patients. Infection 1998;26:359–363.

6 Volpi A, Gross G, Hercogova J, Johnson RW: Current management of herpes zoster: the European view. Am J Clin Dermatol 2005;6:317–325.

7 Schmader K: Herpes zoster in adults. Clin Infect Dis 2001;32:1481–1486.

8 Gross G, Schofer H, Wassilew S, Friese K, Timm A, Guthoff R, Pau HW, Malin JP, Wutzler P, Doerr HW: Herpes zoster guideline of the German Dermatology Society (DDG). J Clin Virol 2003;26: 277–289.

9 Wutzler P, De Clercq E, Wutke K, Farber I: Oral brivudin vs. intravenous acyclovir in the treatment of herpes zoster in immunocompromised patients: a randomized double-blind trial. J Med Virol 1995;46:252–257.

10 Dworkin RH, Perkins FM, Nagasako EM: Prospects for the prevention of postherpetic neuralgia in herpes zoster patients. Clin J Pain 2000;16(suppl):90–100.

11 Feder HM Jr, Hoss DM: Herpes zoster in otherwise healthy children. Pediatr Infect Dis J 2004;23:451–457; quiz 458–460. Review.

12 Santee JA: Corticosteroids for herpes zoster: what do they accomplish? Clin Dermatol 2002;3: 517–524.

13 Wood MJ, Johnson RW, Mc Kendrick MW, Taylor J, Mandal BK, Crooks J: A randomized trial of acyclovir for 7 days or 21 days with and without prednisone for treatment of acute herpes zoster. N Eng J Med 1994;330:896–900.

Gerd Gross, MD
Professor of Dermatology and Venereology, Director
Department of Dermatology and Venereology
University of Rostock
Augustenstrasse 80–84
DE–18055 Rostock (Germany)
Tel. +49 381 4949700, Fax +49 381 4949702, E-Mail gerd.gross@med.uni-rostock.de

Gross G, Doerr HW (eds): Herpes Zoster.
Monogr Virol. Basel, Karger, 2006, vol 26, pp 123–130

Antiviral Therapy of Shingles in Dermatology

H. Martina Lilie, Sawko W. Wassilew

Dermatology Department, Klinikum Krefeld, Krefeld, Germany

The major goals of therapy in patients with herpes zoster are:
(1) shortening of virus replication,
(2) prevention of dissemination of skin lesions,
(3) reduction of acute zoster-associated pain (ZAP),
(4) prevention of chronic ZAP and postherpetic neuralgia (PHN),
(5) prevention of other complications, e.g. ophthalmic involvement.

Recent research has shown that antiviral therapy with acyclovir, valacyclovir, famciclovir, and brivudin, started as early as possible, can significantly shorten viral replication, prevent lesion dissemination and reduce intensity and duration of ZAP particularly in elderly patients, provided that treatment is started early in the course of disease. This suggests that antiviral therapy (table 1) should be offered to all patients as soon as herpes zoster is diagnosed, preferably within 72 h after onset of rash. In patients of any age with ophthalmic herpes zoster and in all immunocompromised patients antiviral therapy should be started even later as long as viral replication can be considered in skin and nerves, e.g. as long as new blisters appear in the skin.

Dermatologists are trained to diagnose early skin lesions as herpes zoster and should be consulted in time.

In the immunocompetent patient antiviral therapy should normally be given for 7 days, whereas in the immunocompromised host antiviral therapy should be given intravenously and for at least 10 days.

Immunocompetent Patients

Acyclovir

Acyclovir (9-{2-hydroxyethoxymethyl}guanine) is a nucleoside analog. After penetrating into the infected cells acyclovir is converted to its monophosphate

Table 1. Antiviral therapy in immunocompetent patients with herpes zoster: management with antivirals, therapy for 7 days [30, 31]

Acyclovir, i.v.	5–10 mg/kg body-weight	3×/day	world
Acyclovir, or.	800 mg	5×/day	world
Famciclovir			
or.	250 mg	3×/day	Europe
or.	500 mg	3×/day	USA
or.	750 mg[1]	1×/day	UK
Valacyclovir, or.	1,000 mg	3×/day	world
Brivudin, or.	125 mg	1×/day	Europe

[1]Data only on acute pain.

derivative by the viral thymidine kinase and to acyclovir triphosphate by cellular kinases. The resulting acyclovir triphosphate inhibits viral DNA polymerase as a competitive inhibitor of guanosine triphosphate [1].

Acyclovir is available in intravenous and oral preparations for the management of herpes zoster. The efficacy and safety of acyclovir has been demonstrated in several placebo-controlled trials. When given intravenously, acyclovir was found to reduce acute pain and to shorten cutaneous healing [2]. Acyclovir 800 mg 5 times daily for 7 days speeds skin healing and reduces the severity of acute neuritis. The benefit of acyclovir therapy was most evident when therapy started within 72 h of disease onset [3–9]. A meta-analysis of the acyclovir data showed that acyclovir has a significant benefit in the reduction of ZAP since it shortened the time to complete cessation of pain in all patients by nearly 80% compared with placebo [10]. Other studies have shown no benefit of acyclovir in reducing the duration of PHN [11, 12]. Typical side effects are transient rises in serum creatinine or urea, nausea, vomiting, diarrhoea, stomach pain, rash and headache.

If given orally, the mean steady-state plasma level of acyclovir is very low, whereas much higher plasma levels can be reached by intravenous application of acyclovir. Because of its poor bioavailability of about 20% and its short intracellular half-life, ester derivatives of antiviral agents, the so-called prodrugs, were developed, which have much better pharmacocinetic properties. These substances are valacyclovir (prodrug of acyclovir) and famciclovir (prodrug of penciclovir).

Valacyclovir

Valacyclovir, the l-valyl ester of acyclovir, is rapidly converted to acyclovir. This results in a significantly (3–5-fold) increased oral acyclovir bioavailability of about 65%. It is FDA approved since 1995. The benefit of orally administered valacyclovir 1,000 mg three times daily has been shown in several large placebo-controlled trials [13, 14]. Skin healing occurred at a similar time with valacyclovir and acyclovir, but cessation of ZAP occurred 34% faster in the valacyclovir group (HR = 1.34, 95% CI: 1.12, 1.60). Patients treated with valacyclovir had only a 19% incidence of pain at 6 months in comparison to those treated with acyclovir with an incidence of 26%, $p = 0.02$ [13]. In another randomized study [15] on 110 patients, valacyclovir reduced the incidence of the ocular complications of herpes zoster ophthalmicus to a similar degree as acyclovir 800 mg five times daily. Overall, valacyclovir therapy has a positive effect on the reduction of ZAP and the prevention or the reduction of ocular complications [3, 15, 16]. Moreover, valacyclovir was well-tolerated, its safety profile being similar to that of acyclovir. Therefore, valacyclovir appears to be more suitable than oral acyclovir, especially because of its simple three-times-daily dose regimen.

Famciclovir

Famciclovir is the prodrug of penciclovir, a guanine nucleoside analog, and was approved by the FDA in 1994. The substance is rapidly absorbed and converted to penciclovir during transit through the intestinal wall and liver. Afterwards penciclovir is converted to its monophosphate by the VZV thymidine kinase within the VZV-infected cells. The monophosphate is subsequently converted by cellular kinases to the diphosphate and triphosphate forms. Penciclovir triphosphate has a significantly longer intracellular half-life than acyclovir triphosphate. The advantage of famciclovir is its good oral bioavailability of about 77% while being administered less frequently than acyclovir [3, 16].

Data from clinical studies did not advocate one particular, universally accepted dose regimen of famciclovir. Therefore, the approved doses vary in different countries: 500 mg t.i.d. especially in the USA, 250 mg t.i.d. in some European countries, and 750 mg q.d. e.g. in the United Kingdom (data only on acute pain). In several large placebo-controlled trials, the benefit of famciclovir could be demonstrated in the resolution of skin lesions and in accelerating the resolution of acute neuritis [3, 14, 16, 17]. Famciclovir is equally effective as valacyclovir. No differences could be shown between famciclovir and valacyclovir concerning rash healing, resolution of acute pain and PHN [14]. Famciclovir is as effective as oral acyclovir 5×800 mg daily for herpes zoster ophthalmicus concerning prevention of ocular complications [18].

Brivudin

Brivudin, [(e)-5-(2-bromovinyl)-2′deoxyuridine], is a potent virostatic agent with an exceptionally high and selective activity against VZV and herpes simplex virus type 1. De Clercq and co-workers demonstrated that brivudin has a greater antiviral in vitro activity against varicella–zoster virus than acyclovir and penciclovir [19–22].

Pharmacocinetic data showed that brivudin has a bioavailability of approximately 30%, presumably due to first-pass metabolism. The substance has a long plasma half-life of approximately 16 h (>95% bound to plasma proteins). The antiviral activity of brivudin depends on its phosphorylation by viral, followed by cellular enzymes and the ensuing interaction with the viral enzymes, but not cellular DNA polymerase. Brivudin undergoes hydrolysis to bromovinyl uracil by pyrimidine nucleoside phosphorylase, mainly in the liver. The second phosphorylation step, however, is catalyzed by the viral thymidine kinase.

The efficacy of brivudin has been shown in two large randomized, double blind multicenter studies, one on a total of 1,227 patients comparing brivudin 125 mg once daily with standard acyclovir 800 mg five times daily for 7 days [23–25], and the other on a total of 2,025 patients comparing brivudin 125 mg once daily with famciclovir 250 mg t.i.d. for 7 days [26].

Compared with acyclovir in the standard dose of 800 mg five times daily, brivudin 125 mg once daily for 7 days showed a significantly better antiviral activity (faster stop of viral replication), and it was as effective as acyclovir with respect to the resolution of acute zoster pain. With its once daily dosing schedule, brivudin offers a considerable advantage over current antiviral therapies. It is as yet available in Germany and had been licensed in many other European countries like Italy, Greece, Spain, Austria, Portugal and Luxemburg. It could be demonstrated that brivudin is well-tolerated and has a similar safety profile as acyclovir and famciclovir. The most frequently reported side effects were headache and gastrointestinal complaints, similar to the known side effects of acyclovir and famciclovir [16, 23, 24, 26, 27].

In conclusion, brivudin 125 mg once daily is as effective as famciclovir 250 mg t.i.d. in reducing the prevalence and the duration of ZAP and PHN 3 months after start of therapy. Although the duration of PHN was shorter in the brivudin group than in the famciclovir group, this difference was not statistically significant [26].

In animal trials, the toxicity of brivudin was low. In long-term (2-year) animal carcinogenicity assays, no tumors were observed in mice, whereas tumors in testes and the liver were observed in rats after a 2-year therapy with an extremely high brivudin dose of 30 mg/kg body-weight/day. In lower doses of up to 12 mg/kg body-weight/day, no tumors were observed. As the established dose in man (125 mg once daily) equals to 1.8 mg/kg body-weight/day and as it is given for only 7 days, carcinogenicity observed in the animal after long-term use does not appear predictive of human risk. Brivudin has no mutagenic potency and did not induce cell-transformation [28, 29].

Table 2. Antiviral therapy in immunocompromised patients with herpes zoster: management with antivirals, therapy for 7–10 days

Acyclovir, i.v.	10 mg/kg body-weight	3×/day
Foscarnet, i.v.	120–200 mg (in 2–3 doses)	/day
Acyclovir, or.	800 mg	5×/day
Valacyclovir, or.	1,000 mg	3×/day
Famciclovir, or.	500 mg	3×/day

As the metabolite bromovinyl uracil interferes with the metabolism of 5-fluorouracil through the inhibition of the enzyme dihydropyrimidine dehydrogenase, brivudin and 5-FU are not allowed to be administered together [27].

Studies using treatment with famciclovir or brivudin for herpes zoster oticus and for herpes zoster ophthalmicus have not yet been published.

Immunocompromised Host

Immunocompromised patients such as transplant recipients, patients with malignancies, HIV-infection and others, receiving immunosuppressant therapy, corticosteroids and chemotherapy are at greater risk of herpes zoster and its complications.

Normally intravenous acyclovir therapy is recommended for the immunocompromised patient [27, 32], but oral antivirals can also be considered in some cases, table 2 [33]. In immunocompromised patients with suspected acyclovir-resistent VZV foscarnet can be administered [1, 3, 34]. Foscarnet is a pyrophosphate analog of phosphonoacetic acid. The substance inhibits the DNA polymerase by directly blocking the pyrophosphate binding site [1]. Due to the less oral bioavailability the substance has to be given intravenously with infusion of 120–200 mg/kg/day in 2–3 doses unless first symptoms of renal failure appear [1]. In cases, where foscarnet may also be ineffective due to gene mutations intravenous cidofovir is the only alternative treatment [35].

Brivudin is not admitted for the treatment in immunocompromised patients.

In summary, the effects of all virostatics given orally or intravenously are comparable concerning the resolution of virus replication, cessation of dissemination of skin lesions and reduction of acute herpes zoster pain. Concerning the incidence and/or prevalence of chronic pain, ZAP and PHN valacyclovir, famciclovir and brivudin in different dosages are comparably effective. Normally all antivirals are well-tolerated, but transient side effects such as headache, gastrointestinal and neurological complaints are possible in all antiviral drugs [31].

Adjunctive Therapies

During the acute phase of herpes zoster, cool compresses or drying solutions, creams, and ointments can help to alleviate local symptoms and dry the skin lesions. They have no proven effect on shortening the disease or preventing complications and PHN.

All virostatics have comparable but not sufficient influence on pain. Therefore, concomitant analgesic therapy is recommended and consists of a variety of modalities including analgesics, narcotics [36], early use of tricyclic antidepressants (amitriptyline) in elderly patients [37–39], opioids [40–42], gabapentin [43–45], capsaicin [5, 46], local 5%-lidocaine-patch [46], cutaneous stimulation, sympathectomy/nerve blocks [47, 48] and corticosteroids therapy. Corticosteroids should be combined with an antiviral therapy in patients ≥50 years of age only with significant acute pain at presentation, if they have no contraindications for high dose corticosteroids administration, i.e. high blood pressure, diabetes, etc. Benefits of this adjunctive therapy are improvement in quality of life, total cessation of analgesic use and undisturbed sleep at night. The dosage of prednisone is recommended at least 1 mg/kg body-weight and day respectively 60 mg daily for first week, 30 mg daily the second week and 15 mg daily the third week. Nevertheless, the adjunctive corticosteroid therapy has been considered controversial, especially due to the fact that no benefit could be proven upon preventing ZAP, PHN or ocular complications [1, 36, 49–51].

References

1 Whitley RJ: Approaches to the treatment of varicella-zoster virus infections in varicella-zoster virus, molecular biology, pathogenesis and clinical aspects. Contrib Microbiol 1999;3:158–172.

2 McGill J, MacDonald DR, Fall C, McKendrick GD, Copplestone A: Intravenous acyclovir in acute herpes zoster infection. J Infect 1983;6:157–161.

3 Balfour HH Jr: Antiviral drugs. N Engl J Med 1999;340:1255–1268.

4 Crooks RJ, Jones DA, Fiddian AP: Zoster-associated chronic pain: an overview of clinical trial with acyclovir. Scand J Infect 1991;80(suppl):62–68.

5 Dworkin RH, Carrington D, Cunningham A, Kost RG, Levin MJ, McKendrick MW, Oxman MN, Rentier B, Schmader KE, Tappeiner G, Wassilew SW, Whitley RJ: Assessment of pain in herpes zoster: lessons learned from antiviral trials. Antiviral Res 1997;33:73–85.

6 Harding SP, Porter SM: Oral acyclovir in herpes zoster ophthalmicus. Curr Eye Res 1991;10: 177–182.

7 Heidl M, Scholz H, Doerffel W, Herrmann J, Wutzler P: Vergleichende Pruefung der Wirksamkeit von Bromvinyldesoxyuridin und Aciclovir bei Varicella-Zoster-Virus-Infektionen immunsupprimierter Kinder – eine prospektive, randomisierte Studie. Z Klin Med 1990;45:1259–1262.

8 Huff JC, Drucker JL, Clemmer A, Laskin OL, Connor JD, Bryson YL, Balfour HH Jr: Efficacy of oral acyclovir on pain resolution in herpes zoster: a reanalysis. J Med Virol 1993;1(suppl): 93–96.

9 Morton P, Thomson AN: Oral acyclovir in the treatment of herpes zoster in general practice. N Z Med J 1989;102:93–95.

10 Wood MJ, Kay R, Dworkin RH, Soong SJ, Whitley RJ: Oral acyclovir therapy accelerates pain resolution in patients with herpes zoster: a meta-analysis of placebo-controlled trials. Clin Infect Dis 1996;22:341–347.
11 McKendrick MW, Care C, Burke C, Hickmott E, McKendrick GD: Oral acyclovir in herpes zoster. J Antimicrob Chemother 1984;14:661–665.
12 van den Broek PJ, van der Meer JW, Mulder JD Versteeg J, Mattie H: Limited value to acyclovir in the treatment of uncomplicated herpes zoster: a placebo-controlled study. Infection 1984;12: 338–341.
13 Beutner KR, Friedman DJ, Forszpaniak C, Andersen PL, Wood MJ: Valaciclovir compared with acyclovir for improved therapy for herpes zoster in immunocompetent adults. Antimicrob Agents Chemother 1995;39:1547–1553.
14 Tyring SK, Beutner KR, Tucker BA, Anderson WC, Crooks RJ: Antiviral therapy for herpes zoster: randomized, controlled clinical trial of valaciclovir and famciclovir therapy in immunocompetent patients 50 years and older. Arch Fam Med 2000;9:863–869.
15 Colin J, Prisant O, Cochener B, Lescale O, Rolland B, Hoang-Xuan T: Comparison of the efficacy and safety of valaciclovir and acyclovir for the treatment of herpes zoster ophthalmicus. Ophthalmology 2000;107:1507–1511.
16 Wutzler P, Doerr HW: Antivirale Therapie des Zoster. Dtsch Aerztbl 1998;95:79–81.
17 Dworkin RH, Boon RJ, Griffin DR, Phung D: Postherpetic neuralgia: impact of famciclovir, age, rash severity, and acute pain in herpes zoster patients. J Infect Dis1998;178(suppl 1):76–80.
18 Tyring S, Engst R, Corriveau C, Robillard N, Trottier S, Van Slycken S, Crann RA, Locke, LA, Saltzman R, Palestine AG: Famciclovir for ophthalmic zoster: a randomised aciclovir controlled study. Br J Ophthalmol 2001;85:576–581.
19 Andrei G, Snoeck R, Reymen D, Liesnard C: Comparative activity of selected antiviral compounds against clinical strains of varicella zoster virus. Eur J Clin Microbiol Infect Dis 1995;14: 318–329.
20 De Clercq E, Descamps J, De Somer P, Barr PJ, Jones AS, Walker RT: (E)-5-(2-Bromovinyl)-2′-deoxyuridine: a potent and selective anti-herpes agent Proc Natl Acad Sci USA 1979;76: 2947–2951.
21 De Clercq E, Degreef H, Wildiers J, De Jonge G, Drochmans A, Descamps J, De Somer P: Oral (E)-5-(2-bromovinyl)-2′-deoxyuridine in severe Herpes zoster. Brit Med J 1980;281:1178.
22 De Clercq E, Descamps J, Verhelst G, Walker RT, Jones AS, Torrence PF, Shugar D: Comparative efficacy of antiherpes drugs against different strains of Herpes simplex virus. J Infect Dis 1980;141:563–574.
23 Wassilew SW, Wutzler P: Oral brivudin in comparison with acyclovir for improved therapy of herpes zoster in immunocompetent patients: results of a randomized, double-blind, multicentered study. Antiviral Res 2003;59:49–56.
24 Wassilew SW, Wutzler P: Oral brivudin in comparison with acyclovir for herpes zoster: a survey study on postherpetic neuralgia. Antiviral Res 2003;59:57–60.
25 Wassilew SW: Virus- und Schmerzbekämpfung sind oberste Therapieziele. Pharm Ztg 2001;146: 43–49.
26 Wassilew SW, Stubinski BM, Koch I, Schumacher K, Städtler G: Brivudin compared to famciclovir in the treatment of herpes zoster: effect on acute disease and chronic pain in immunocompetent patients – a randomized multinational study. J Europ Acad Dermatol Venereol 2005;19: 47–55.
27 Wutzler P, De Clercq E, Wutke K, Faerber I: Oral brivudin vs. intravenous acyclovir in the treatment of Herpes zoster in immunocompromised patients: a randomised double-blind trial. J Med Virol 1995;46:252–257.
28 De Clercq E, Cassiman JJ: Mutagenic potential of anti-herpes agent. Life Science 1986;38: 281–289.
29 Wutzler P, Thust R: Genetic risks of antiviral nucleoside analogues – a survey. Antiviral Res 2001;49:55–74.
30 Gross G, Schöfer H, Wassilew SW, Friese K, Timm A, Guthoff R, Pau HW, Malin JP, Wutzler P, Doerr HW: Herpes zoster guideline of the German Dermatology Society (DDG). J Clin Virol 2003;26:277–289.
31 Lilie HM, Wassilew SW: The role of antivirals in the management of neuropathic pain in the older patient with herpes zoster. Drugs Aging 2003;20:561–570.

32 Balfour HH Jr, Bean B, Laskin OL, Ambinder RF, Meyers JD, Wade JC, Zaia JA, Aeppli D, Kirk LE, Segreti AC, Keeney RE: Acyclovir halts progression of herpes zoster in immunocompromised patients. J Antimicrob Chemother 1983;12(suppl B):169–179.
33 Tyring S, Belanger R, Bezwoda W, Ljungman P, Boon R, Saltzman RL, Colaborative Famciclovir Immunocompromised Study Group: A randomized, double-blind trial of famciclovir versus acyclovir for the treatment of localized dermatomal herpes zoster in immunocompromised patients. Caner Invest 2001;19:13–22.
34 Breton G, Fillet AM, Katlama C, Bricaire F, Caumes E: Acyclovir-resistant herpes zoster in human immunodeficiency virus-infected patients: results of foscarnet therapy. Clin Infect Dis 1998;27: 1525–1527.
35 Safrin S, Cherrington J, Jaffe HS: Clinical uses of cidovofir. Rev Med Virol 1997;7:145–156.
36 Pasqualucci A, Pasqualucci V, Galla F, De Angelis V, Marzocchi V, Colussi R, Paoletti F, Girardis M, Lugano M, Del Sindaco F: Prevention of post-herpetic neuralgia: acyclovir and prednisolone versus epidural local anesthetic and methylprednisolone. Acta Anaesthesiol Scand 2000;44:910–918.
37 Bowsher D: The effects of pre-emptive treatment of postherpetic neuralgia with amitriptyline: a randomized, double-blind, placebo-controlled trial. J Pain Symptom Manage 1997;13:327–331.
38 Graff-Radford SB, Shaw LR, Naliboff BN: Amitriptyline and fluphenazine in the treatment of postherpetic neuralgia. Clin J Pain 2000;16:188–192.
39 Watson CPN, Vernich L, Chipman M, Reed K: Nortriptyline versus amitriptyline in postherpetic neuralgia: a randomized trial. Neurology 1998;51:1166–1171.
40 Dühmke RM, Cornblath DD, Hollinshead JRF: Tramadol for neuropathic pain (Cochrane Review); in The Cochrane Library, Issue 3, 2004.
41 Dworkin RH, Schmader KE: Treatment and prevention of postherpetic neuralgia. Clin Infect Dis 2003;36:877–882.
42 Rowbotham MC, Twilling L, Davies PS, Reisner L, Taylor K, Mohr D: Oral opioid therapy for chronic periphal and central neuropathic pain. N Engl J Med 2003;348:1223–1232.
43 Curran MP, Wagstaff AJ: Gabapentin in postherpetic neuralgia. CNS Drugs 2003;17:975–982.
44 Rice ASC, Morton S, Postherpetic Neuralgia Study Group: Gabapentin in postherpetic neuralgia: a randomised double blind, placebo controlled study. Pain 2001;94:215–224.
45 Rowbotham M, Harden N, Stacey B, Bernstein P, Magnus-Miller L, for the Gabapentin Postherpetic Neuralgia Study Group: Gabapentin for the treatment of postherpetic neuralgia: a randomized controlled trial. JAMA 1998;280:1837–1842.
46 Galer BS: Advances in the Treatment of Postherpetic Neuralgia: the topical lidocaine patch. Todays Ther Trends 2000;18:1–20.
47 Winnie AP, Hartwell: Relationship between time of treatment of acute herpes zoster with sympathetic blockade and prevention of post-herpetic neuralgia: clinical support for a new theory of the mechanism by which sympathetic blockade provides therapeutic benefit. Reg Anesth 1993;18:277–282.
48 Wu CL, Marsh A, Dworkin RH: The role of sympathetic nerve blocks in herpes zoster and postherpetic neuralgia. Pain 2000;87:121–129.
49 Gnann JW, Whitley RJ: Herpes zoster. N Eng J Med 2002;347:340–346.
50 Whitley RJ, Weiss H, Gnann JW Jr, Tyring S, Mertz GJ, Pappas PG, Schleupner CJ, Hayden F, Wolf J, Soong SJ: Acyclovir with and without prednisone for the treatment of herpes zoster. A randomized, placebo-controlled trial. The National Institute of Allergy and Infectious Diseases Collaborative Antiviral Study Group. Ann Intern Med 1996;125:376–383.
51 Wood MJ, Johnson RW, McKendrick MW, Taylor J, Mandal BK, Crooks J: A randomized trial of acyclovir for 7 days or 21 days with and without prednisolone for treatment of acute herpes zoster. N Engl J Med 1994;330:896–900.

Dr. H. Martina Lilie
Dermatology Department, Klinikum Krefeld
Lutherplatz 40
DE–47805 Krefeld (Germany)
Tel. +49 2151 32 2880, Fax +49 2151 32 2005, E-Mail lilie@klinikum-krefeld.de

Gross G, Doerr HW (eds): Herpes Zoster.
Monogr Virol. Basel, Karger, 2006, vol 26, pp 131–142

Highly Potent and Selective Inhibition of Varicella–Zoster Virus Replication by Bicyclic Furo[2,3-*d*]pyrimidine Nucleoside Analogues (BCNAs)

Erik De Clercq

Rega Institute for Medical Research, Katholieke Universiteit Leuven, Leuven, Belgium

Varicella-zoster virus (VZV) causes two distinct syndromes [1]: primary infection presents as varicella (or chickenpox), a usually benign illness during childhood, and subsequent reactivation of latent VZV in dorsal-root ganglia results in a localized cutaneous eruption termed 'herpes zoster' (or 'shingles'). The annual incidence of herpes zoster is 0.2%, although in the elderly (persons older than 75 years) and in immunocompromised patients it can exceed 1 and 2%, respectively [1]. Therapy for herpes zoster is warranted to accelerate healing, limit the severity and duration of acute and chronic pain, and reduce complications (among which postherpetic neuralgia, defined as pain that persists for more than 30 days after the onset of rash, is the most feared complication in immunocompetent patients); in immunocompromised patients, an additional therapeutic objective is to reduce the risk of dissemination of VZV [1].

In the United States three drugs – acyclovir (ACV), valacyclovir and famciclovir – are approved for the treatment of herpes zoster: ACV at a dosage of 800 mg every 4 h (5 times daily) for 7–10 days, valacyclovir at a dosage of 1,000 mg every 8 h (3 times daily) for 7 days, and famciclovir at a dosage of 500 mg every 8 h (3 times daily) for 7 days: because of the simpler dosing regimens, valacyclovir and famciclovir (which are therapeutically equipotent) are preferred to ACV for the treatment of herpes zoster [1]. In addition to ACV, valacyclovir and famciclovir, also brivudin (BVDU) has been licensed for the therapy of herpes zoster in Germany and several other European Countries (Austria, Belgium, Greece, Italy, Luxemburg, Portugal, Spain): BVDU has the

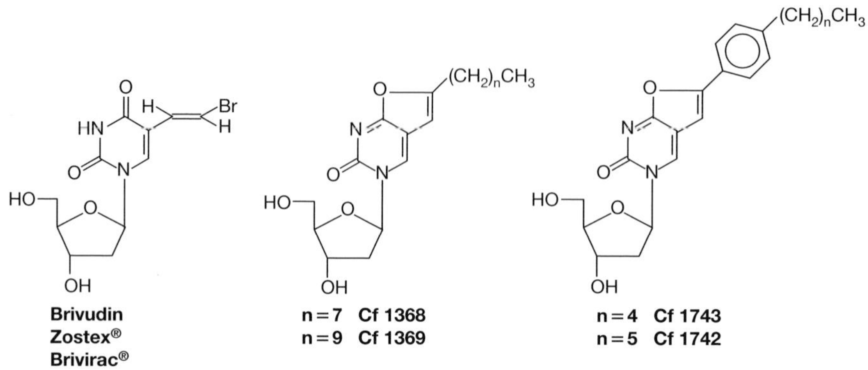

Fig. 1. Structures of BVDU and prototype furo[2,3-*d*]pyrimidine nucleoside analogues: Cf 1368, Cf 1369, Cf 1742 and Cf 1743.

simplest dosing regimen of all, as it must be administered (orally) only once daily (125 mg) for 7 days, to achieve equivalent efficacy to valacyclovir and famciclovir [2].

Here, I report on a totally new class of compounds, that of the bicyclic furo[2,3-*d*]pyrimidine nucleoside analogues (BCNAs), which can be viewed as somewhat related to BVDU (fig. 1), but which differ from BVDU in several biological and biochemical aspects that make these furopyrimidine derivatives ideally suited as selective (candidate) anti-VZV drugs.

Structure–Activity Relationships of Furo[2,3-*d*]Pyrimidine Nucleoside Analogues (BCNAs)

The target structures were originally obtained as by-products in Pd-catalysed coupling of terminal alkynes with 5-iodo-nucleoside analogues such as 5-iodo-2′-deoxyuridine [3]. Among a first series of 6-substituted furopyrimidine derivatives, the 6-octyl-substituted derivative Cf 1368 (fig. 1) was found to inhibit the replication of VZV (strains OKA and YS) in human embryonic lung cells at a 50% effective concentration (EC_{50}, concentration required to inhibit virus-induced plaque formation by 50%) of 0.008 µM, while not being toxic at a concentration (CC_{50}, cytotoxic concentration required to inhibit human embryonic lung cell growth by 50%) of 50 µM ($CC_{50} > 50$ µM) [3], thus achieving a selectivity index (SI, ratio CC_{50} to EC_{50}) of >5,000. Also 6-nonyl- and 6-decyl-substituted derivatives showed comparable antiviral activity. Shorter (i.e. heptyl, hexyl) or longer (i.e. undecyl, dodecyl) side chains resulted in decreased anti-VZV activity (table 1).

Further lead optimization resulted in the synthesis of 6-(*p*-alkylphenyl)-substituted furopyrimidine nucleoside analogues. Among these derivatives, the 6-(*p*-pentylphenyl)- and 6-(*p*-hexylphenyl)-substituted derivatives (Cf 1743 and Cf 1742, respectively) (fig. 1) [EC_{50}: ~0.0001–0.0005 μM and selectivity index (>100,000)] emerged as the most potent and selective anti-VZV compounds ever reported (table 2) [4]. Shortening or lengthening the alkyl chain length of the 6-pentyl- or 6-hexylphenyl chain with one or more carbons again resulted in decreased anti-VZV activity. For both the alkyl and *p*-alkylphenyl-substituted compounds an optimal length of the side chain could be defined; thus, for the alkyl derivatives, this appeared to be 8–10 carbons, and for the *p*-alkylphenyl derivatives 5 or 6 carbons, the phenyl group being equivalent to 3–4 carbons.

Anti-VZV Properties of Bicyclic Furo[2,3-*d*]Pyrimidine Nucleoside Analogues (BCNAs)

From the 6-alkyl-substituted furopyrimidine nucleoside series, the octyl derivative Cf 1368, and from the 6-*p*-alkylphenyl-substituted furopyrimidine nucleoside series, the pentylphenyl derivative Cf 1743 and hexylphenyl derivative Cf 1742 (fig. 1) were selected for further exploration of their anti-VZV properties. As compared to (*E*)-5-(2-bromovinyl)-2′-deoxyuridine (BVDU), Cf 1743 and Cf 1742 were about 10–20-fold more potent against VZV, and, as compared to ACV and penciclovir (PCV), they were even 10,000-fold more potent in their activity against VZV [5–8].

When Cf 1368, Cf 1742 and Cf 1743 were evaluated against a variety of clinical VZV isolates, EC_{50} values were obtained that were quite comparable to those obtained for the laboratory OKA and YS VZV strains [9]. Cf 1368, Cf 1742 and Cf 1743 inhibited the replication of wild-type clinical VZV isolates at a mean EC_{50} value of 0.032, 0.00082 and 0.00044 μM, respectively, whereas for BVDU the mean EC_{50} value was 0.0098 μM. The corresponding EC_{50} values for ACV, PCV and foscarnet (phosphonoformic acid [PFA]) were 3.51, 3.31 and 83.62 μM, respectively (fig. 2).

Thus, in terms of activity against both laboratory and clinical VZV strains, the order of (decreasing) potency was Cf 1743 ~ Cf 1742 > Cf 1368 ~ BVDU > PCV ~ ACV > PFA (fig. 2). Cf 1743 and Cf 1742 are much more potent in their anti-VZV activity than the drugs that are formally approved for the treatment of VZV infections [PCV (penciclovir, oral prodrug form: famciclovir) and ACV (acyclovir, oral prodrug form: valacyclovir)] and than foscar-

Table 1. Anti-VZV activity of BCNAs containing an alkyl side chain on the furanyl ring [3]

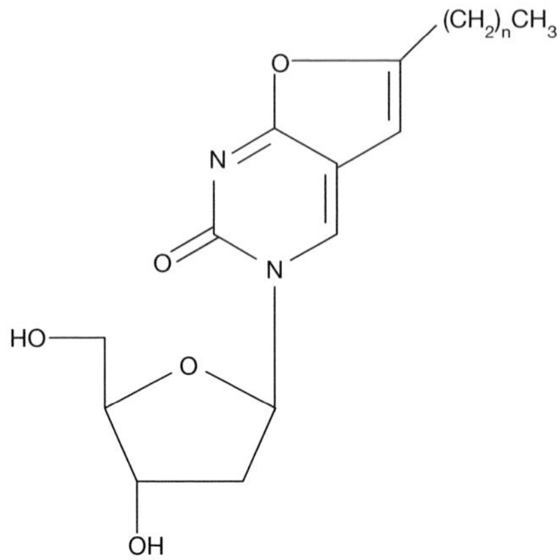

Alkyl side chain	n	EC_{50} (μM)			CC_{50} (μM)
		VZV		VZV TK$^-$	
		OKA	YS	YS-R	
Hexyl	5	1.3	2.8	>50	>200
Heptyl	6	0.12	0.33	>50	>50
Octyl	7 (**Cf 1368**)	0.008	0.024	>50	>50
Nonyl	8	0.02	0.02	>200	>200
Decyl	9 (**Cf 1369**)	0.015	0.008	>50	>50
Undecyl	10	0.4	0.37	>50	>200
Dodecyl	11	0.8	1.2	>50	>200
BVDU		0.005	0.005	>50	>400
Acyclovir		1.9	2.1	>50	>200

net which is occasionally used in the treatment of ACV-resistant VZV infections in immunocompromised patients [10].

In vitro resistance to the furo[2,3-*d*]pyrimidine nucleoside analogues could be readily obtained upon repeated passages of VZV (OKA strain) in the presence of Cf 1368, Cf 1742 and Cf 1743. These resistant strains (Cf 1368^r, Cf 1742^r and Cf 1743^r) also showed cross-resistance to BVDU, BVaraU and ACV, three compounds that for their anti-VZV activity strongly depend on phosphorylation by the VZV-encoded thymidine kinase (TK). Vice versa, the BVDU^r,

Table 2. Anti-VZV activity of furo[2,3-*d*]pyrimidine nucleoside analogues containing a *p*-alkyl-substituted phenyl group as the side chain on the furanyl ring [4]

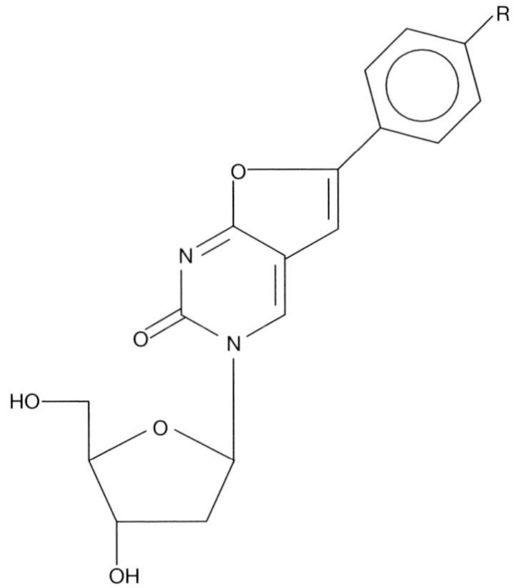

R	EC_{50} (μM)		CC_{50} (μM)
	VZV		
	OKA	YS	
H	0.28	0.16	>200
Methyl	0.06	0.06	>200
Ethyl	0.09	0.07	123
Propyl	0.01	0.008	188
Butyl	0.0022	0.0005	>200
Pentyl (**Cf 1743**)	0.0003	0.0001	>200
Hexyl (**Cf 1742**)	0.0005	0.0001	18
Heptyl	0.0054	0.003	18
Octyl	0.04	0.027	>200

BVaraU^{r} and ACV^{r} mutants also proved cross-resistant to the furopyrimidine nucleoside analogues [9]. These data suggest that the furopyrimidine nucleoside analogues, akin to BVDU, BVaraU and ACV, select for drug-resistant VZV mutants that are deficient in their TK-dependent phosphorylating capacity. On the other hand, PCV^{r}, PFA^{r} and PMEA^{r} strains [9], that were selected for resistance to PCV, foscarnet or adefovir, respectively, retained marked susceptibility

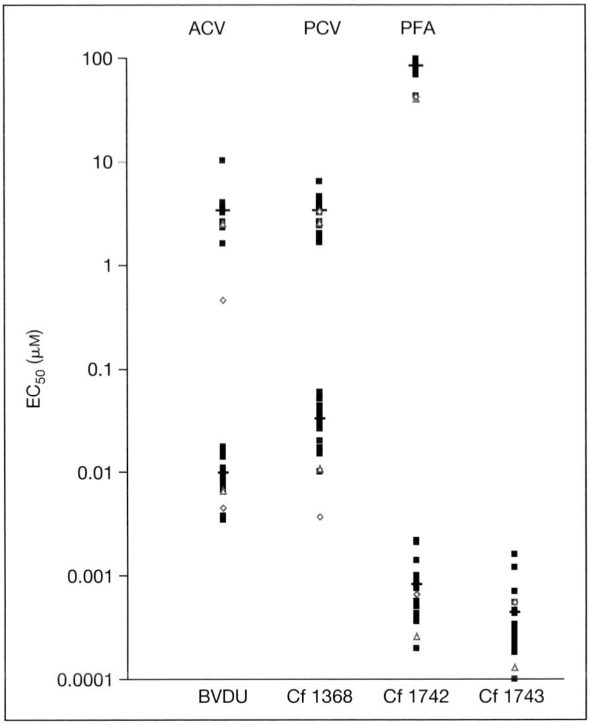

Fig. 2. Activity of BCNAs and anti-VZV reference compounds [BVDU, ACV, PCV and foscarnet (PFA)] against clinical VZV isolates (■) and VZV reference strains Oka (△) and YS (◇). Mean values for clinical isolates are indicated by horizontal bars. Data taken from reference [9].

to Cf 1742 and Cf 1743. This suggests that the PCV, PFA and PMEA resistance phenotype must be attributed to mutations in an enzyme other than the viral TK, most probably the viral DNA polymerase.

Mechanism of Antiviral Action of Bicyclic Furo[2,3-*d*] Pyrimidine Nucleoside Analogues (BCNAs)

All the compounds that are currently used in the treatment of VZV infections (i.e. ACV, PCV, BVDU) are also active against herpes simplex virus type 1 (HSV-1) and, furthermore, ACV and PCV are also active against herpes simplex virus type 2 (HSV-2). The furopyrimidine nucleoside analogues Cf 1368 and Cf 1743, however, are selectively active against VZV (they are not active against HSV-1 or HSV-2 or other herpes viruses such as cytomegalovirus). This is because they are recognized as a substrate for phosphorylation by the VZV-encoded TK [11]. In contrast, the HSV-1 TK, HSV-2 TK as well as the cytosolic TK-1 and mitochondrial TK-2 do not recognize the furopyrimidine nucleoside analogues as substrate. The VZV TK, due to its intrinsic dTMP kinase activity, will successively phosphorylate the furopyrimidine nucleosides to the 5′-monophosphate and 5′-diphosphate form [11]. Whether the compounds are then further phosphorylated to the 5′-triphosphate form could so far not be clearly demonstrated.

Phosphorylation of the furopyrimidine nucleoside analogues by the VZV TK is a prerequisite for their anti-VZV activity, but it clearly is not sufficient a requirement. For a large series of furopyrimidine nucleosides that were examined for both their anti-VZV activity and their inhibitory effects on thymidine phosphorylation by VZV TK (as a parameter of affinity of the compounds to the viral enzyme), no correlation whatsoever was obtained. This means that, while the viral TK is necessary, it is by itself not sufficient to explain the anti-VZV activity of the furopyrimidine nucleoside analogues. How these compounds accomplish their antiviral activity and with which molecular target they eventually interact remain fascinating questions, subject of further investigation in our laboratory.

Metabolic Interactions of Bicyclic Furo[2,3-*d*]Pyrimidine Nucleoside Analogues (BCNAs)

The pyrimidine nucleoside catabolic enzyme, thymidine phosphorylase (TPase), is known to hydrolyse the N-glycosidic bond of various pyrimidine nucleoside analogues, including BVDU [12, 13]. Since the aglycones (free pyrimidine bases) are as such devoid of antiviral activity, the action of TPase may annihilate the antiviral activity of the pyrimidine nucleoside analogues. Interestingly, the furopyrimidine nucleoside analogues Cf 1368 and Cf 1743 were found to be entirely resistant to the phosphorolytic cleavage by both human TPase and bacterial TPase [14].

The free nucleobase of BVDU, (*E*)-5-(2-bromovinyl)uracil (BVU), is an efficient inhibitor of human dihydropyrimidine dehydrogenase (DPD) [15]. DPD is a key enzyme involved in the degradation of natural pyrimidines and pyrimidine analogues such as 5-fluorouracil (5-FU). BVU, through its inhibitory effect on DPD, may significantly potentiate the toxicity of 5-FU. The relevance of this interaction has become dramatically clear in the past, when the simultaneous administration of oral sorivudine (BVaraU) with 5-FU in cancer patients was found to lead to a number of deaths as the result of an accrued 5-FU toxicity [16–18]. The cascade of reactions leading to these fatalities were

(i) phosphorolysis of BVaraU by the intestinal prokaryotic TPase(s), (ii) release of BVU in the plasma, and (iii) inhibition of DPD, with consequently, unacceptably high plasma levels of 5-FU.

In marked contrast with BVU that inhibited human liver DPD activity at an IC_{50} of about 10 μM, the free furo[2,3-*d*]pyrimidine base Cf 1381 (the aglycone of Cf 1368) was completely ineffective in inhibiting DPD at a concentration of 100 or 250 μM [14]. Also, when the furopyrimidine nucleoside analogues Cf 1368 and Cf 1743 were administered to mice in combination with 5-FU, they did not increase the 5-FU plasma levels, while under the same experimental conditions BVU markedly raised the plasma levels of 5-FU [14]. Thus, the furopyrimidine nucleoside analogues may be expected not to influence the 5-FU plasma levels in patients treated with 5-FU for cancer who would concomitantly be treated with the furopyrimidine derivatives for a concurrent VZV infection.

Metabolic Pathways of BVDU and Furo[2,3-*d*]Pyrimidine Nucleoside Analogues

The various enzymatic (both anabolic and catabolic) steps involved in the metabolism of BVDU have been well characterized [19–21]. BVDU is phosphorylated by VZV TK, as well as HSV-1 TK, to the 5′-mono- and 5′-diphosphate, whereupon phosphorylation by cellular enzymes such as nucleoside 5′-diphosphate (NDP) kinase will convert the 5′-di- to the 5′-triphosphate (BVDU-TP) (fig. 3). BVDU-TP represents the active metabolite of BVDU, which interacts with its target enzyme, the viral DNA polymerase, as a competitive inhibitor/alternate substrate with respect to the natural substrate, dTTP. In contrast with BVDU, the bicyclic furo[2,3-*d*]pyrimidine nucleoside analogues are recognized as substrate by the VZV TK only, which converts the compounds successfully to the 5′-mono- and 5′-diphosphate (fig. 4). What happens subsequently is presently unclear, and, also the final target (enzyme) for the antiviral action of the BCNAs is presently unknown.

Although VZV is closely related to the simian varicella virus (SVV), the furo[2,3-*d*]pyrimidine nucleoside analogues are completely inactive against SVV [22]. Yet, the furopyrimidine nucleoside analogues are recognized as substrates by the SVV TK, but, apparently SVV TK-catalysed phosphorylation is necessary but not sufficient for the furopyrimidine nucleoside analogues to display antiviral activity [22].

In contrast with BVDU (fig. 3), the furopyrimidine nucleoside analogues (fig. 4) are not recognized as substrate by TPase which would otherwise cleave off the free aglycone, but even the free aglycone should be formed (hypotheti-

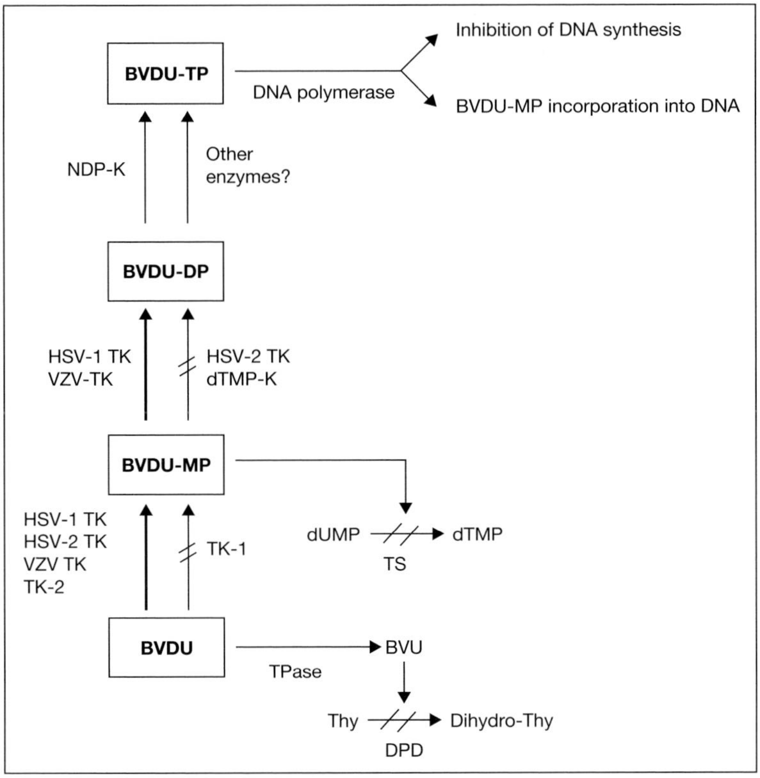

Fig. 3. Anabolic and catabolic steps involved in the metabolism of BVDU.

cally), it would not impose a problem, since the furopyrimidine aglycone has been shown not to interfere with DPD, the enzyme that is responsible for the degradation of thymine, uracil and therapeutically used uracil derivatives (such as the anticancer agent 5-FU).

Recent (unpublished) data also suggest that, unlike BVDU-MP, the furopyrimidine nucleoside monophosphates would not serve as a substrate for 5′-deoxynucleotidases, which, otherwise, may counteract the essential phosphorylation process engendered by the VZV TK [23].

Conclusions

The bicyclic furo[2,3-*d*]pyrimidine nucleoside analogues (BCNAs) are exquisitely potent and specific antiviral agents endowed with an exclusive selectivity for VZV. In contrast with BVDU, the furopyrimidine nucleosides are

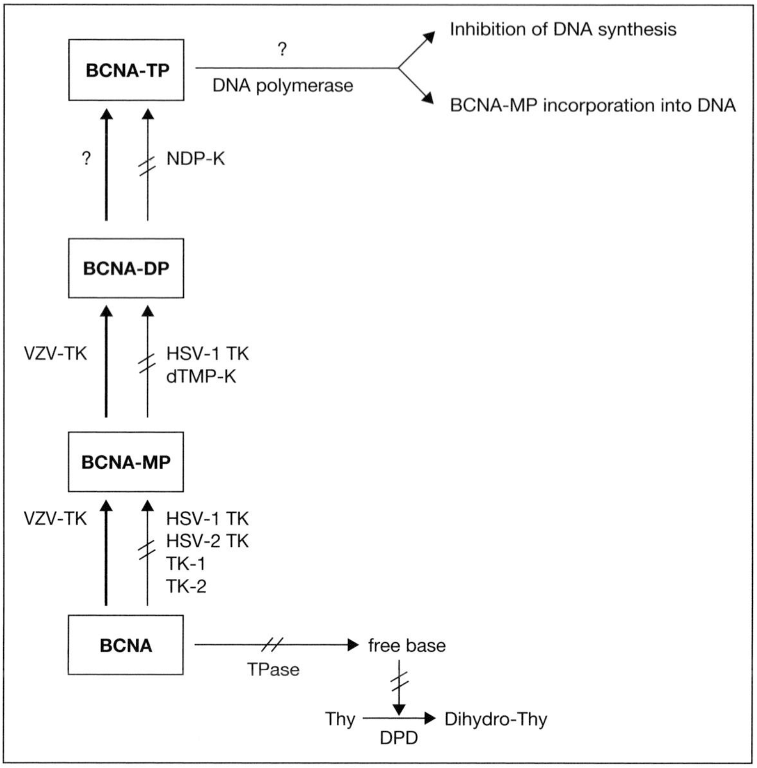

Fig. 4. Anabolic and catabolic steps involved in the metabolism of bicyclic furo[2,3-*d*] pyrimidine nucleoside analogues (BCNAs).

recognized solely by the TK and (therewith associated thymidylate kinase activity) of VZV. They are not hydrolysed by human or bacterial TPases, and their free bases do not inhibit human dihydropyrimidine dehydrogenase. They are also characterized by a number of favourable properties: straightforward chemical synthesis, good stability in biological fluids, virtual absence of cytotoxicity, pronounced oral bioavailability, and no interference with the degradation of 5-FU. The furopyrimidine nucleoside analogues proved equally effective against laboratory VZV strains as well as clinical VZV isolates. Cf 1742 and Cf 1743, the most potent among the furopyrimidine series, inhibited the replication of VZV in cell culture at subnanomolar concentrations. The precise mechanism of antiviral action of the furopyrimidine nucleoside analogues still remains to be clarified. Their unprecedented potency, specificity and selectivity for VZV highly justifies further development of these compounds as candidate drugs for the treatment of VZV infections, i.e. varicella and herpes zoster.

Acknowledgments

This review is based on the original investigations carried out at the Rega Institute (G. Andrei, A. Camps, L. Naesens, R. Sienaert, R. Snoeck, L. van Berckelaer, R. Van Berwaer, L. Vandenheurck, J. Balzarini and E. De Clercq), the University of Wales Cardiff Medicinal Chemistry Group (headed by C. McGuigan), Oxford University (R. Esnouf, D. Stammers), and the Academic Clinic Amsterdam (A. van Kuilenburg). I thank Christiane Callebaut for her proficient editorial assistance.

References

1 Gnann JW Jr, Whitley RJ: Herpes zoster. N Engl J Med 2002;347:340–346.
2 Gross G, Schöfer H, Wassilew S, Friese K, Timm A, Guthoff R, Pau HW, Malin JP, Wutzler P, Doerr HW: Herpes zoster guideline of the German Dermatology Society (DDG). J Clin Virol 2003;26:277–289.
3 McGuigan C, Yarnold CJ, Jones G, Velázquez S, Barucki H, Brancale A, Andrei G, Snoeck R, De Clercq E, Balzarini J: Potent and selective inhibition of varicella-zoster virus (VZV) by nucleoside analogues with an unusual bicyclic base. J Med Chem 1999;42:4479–4484.
4 McGuigan C, Barucki H, Blewett S, Carangio A, Erichsen JT, Andrei G, Snoeck R, De Clercq E, Balzarini J: Highly potent and selective inhibition of varicella-zoster virus by bicyclic furopyrimidine nucleosides bearing an aryl side chain. J Med Chem 2000;43:4993–4997.
5 De Clercq E, Degreef H, Wildiers J, de Jonge G, Drochmans A, Descamps J, De Somer P: Oral (E)-5-(2-bromovinyl)-2′-deoxyuridine in severe herpes zoster. Br Med J 1980;281:1178.
6 Shigeta S, Yokota T, Iwabuchi T, Baba M, Konno K, Ogata M, De Clercq E: Comparative efficacy of antiherpes drugs against various strains of varicella-zoster virus. J Infect Dis 1983;147:576–584.
7 De Clercq E: The antiviral spectrum of (E)-5-(2-bromovinyl)-2′-deoxyuridine. J Antimicrob Chemother 1984;14(suppl A):85–95.
8 Andrei G, Snoeck R, Reymen D, Liesnard D, Goubau P, Desmyter J, De Clercq E: Comparative activity of selected antiviral compounds against clinical isolates of varicella-zoster virus. Eur J Clin Microbiol Infect Dis 1995;14:318–328.
9 Andrei G, Sienaert R, Naesens L, McGuigan C, De Clercq E, Balzarini J, Snoeck R: Susceptibilities of several clinical varicella-zoster virus (VZV) isolates and drug-resistant VZV strains to bicyclic furano pyrimidine nucleosides. Antimicrob Agents Chemother 2005;49:1081–1086.
10 Safrin S, Berger TG, Gilson I, Wolfe PR, Wofsy CB, Mills J, Biron KK: Foscarnet therapy in five patients with AIDS and acyclovir-resistant varicella-zoster virus infection. Ann Intern Med 1991;115:19–21.
11 Sienaert R, Naesens L, Brancale A, De Clercq E, McGuigan C, Balzarini J: Specific recognition of the bicyclic pyrimidine nucleoside analogs, a new class of highly potent and selective inhibitors of varicella-zoster virus (VZV), by the VZV-encoded thymidine kinase. Mol Pharmacol 2002;61:249–254.
12 Desgranges C, Razaka G, Rabaud M, Bricaud H, Balzarini J, De Clercq E: Phosphorolysis of (E)-5-(2-bromovinyl)-2′-deoxyuridine (BVDU) and other 5-substituted 2′-deoxyuridines by purified human thymidine phosphorylase and intact blood platelets. Biochem Pharmacol 1983;32:3583–3590.
13 Liermann B, Hermann G: (E)-5-(2-Bromovinyl)-2′-deoxyuridine: a good substrate for mammalian pyrimidine nucleoside phosphorylases. Biomed Biochim Acta 1983;42:K35–K38.
14 Balzarini J, Sienaert R, Liekens S, Van Kuilenburg A, Carangio A, Esnouf R, De Clercq E, McGuigan C: Lack of susceptibility of bicyclic nucleoside analogs, highly potent inhibitors of varicella-zoster virus, to the catabolic action of thymidine phosphorylase and dihydropyrimidine dehydrogenase. Mol Pharmacol 2002;61:1140–1145.

15 Desgranges C, Razaka G, De Clercq E, Herdewijn P, Balzarini J, Drouillet F, Bricaud H: Effect of (E)-5-(2-bromovinyl)uracil on the catabolism and antitumor activity of 5-fluorouracil in rats and leukemic mice. Cancer Res 1986;46:1094–1101.
16 Diasio RB: Sorivudine and 5-fluorouracil: a clinically significant drug-drug interaction due to inhibition of dihydropyrimidine dehydrogenase. Br J Clin Pharmacol 1998;46:1–4.
17 Nakayama H, Kinouchi T, Kataoka K, Akimoto S, Matsuda Y, Ohnishi Y: Intestinal anaerobic bacteria hydrolyse sorivudine, producing the high blood concentration of 5-(E)-5-(2-bromovinyl)uracil that increases the level and toxicity of 5-fluorouracil. Pharmacogenetics 1997;7:5–43.
18 Okuda H, Ogura K, Kato A, Takubo H, Watabe T: A possible mechanism of eighteen patient deaths caused by interactions of sorivudine, a new antiviral drug, with oral 5-fluorouracil prodrugs. J Pharmacol Exp Ther 1998;287:791–799.
19 De Clercq E, Walker RT: Chemotherapeutic agents for herpesvirus infections; in Ellis GP, West GB (eds): Progress in Medicinal Chemistry, vol 23. Amsterdam, The Netherlands, Elsevier Science Publishers B.V. (Biomedical Division), 1986, pp 187–218.
20 De Clercq E: Towards a selective chemotherapy of virus infections. Development of bromovinyldeoxyuridine as a highly potent and selective anti-herpetic drug. Verh K Acad Geneesk Belg 1986;48:261–290.
21 De Clercq E: Antiviral activity of nucleoside analogues: the BVDU connection; in Chu D (ed): Recent Advances in Nucleosides: Chemistry and Chemotherapy. New York, NY, USA, Elsevier Science B.V., 2002, pp 433–454.
22 Sienaert R, Andrei G, Snoeck R, De Clercq E, McGuigan C, Balzarini J: Inactivity of the bicyclic pyrimidine nucleoside analogues against simian varicella virus (SVV) does not correlate with their substrate activity for SVV-encoded thymidine kinase. Biochem Biophys Res Commun 2004;315:877–883.
23 De Clercq E: Highly potent and selective inhibition of varicella-zoster virus replication by bicyclic furo[2,3-*d*]pyrimidine nucleoside analogues. Med Res Rev 2003;23:253–274.

Erik De Clercq, MD
Rega Institute for Medical Research
Minderbroedersstraat 10
BE–3000 Leuven (Belgium)
Tel. +32 16 337367, Fax +32 16 337340, E-Mail erik.declercq@rega.kuleuven.ac.be

Gross G, Doerr HW (eds): Herpes Zoster.
Monogr Virol. Basel, Karger, 2006, vol 26, pp 143–153

Therapy of Zoster Pain, Postherpetic Neuralgia and Other Neurological Complications

Ralf Baron

Neurological Clinic, Pain Research and Therapy, University Kiel, Kiel, Germany

Treatment of Acute Herpes Zoster

In most case acute herpes zoster is a self-limiting disease and will resolve without complications with time. Analgesic drugs, i.e. NSAIDS and weak opioids or in severe cases strong opioids, in combination with local antiinflammatory ointments are used to establish adequate control of acute pain during healing of the rash. Furthermore, antiviral therapy (e.g. acyclovir, famciclovir, valacyclovir and brivudin) should be initiated as soon as possible, especially in older patients. If neurological complications, i.e. motor paresis, clinical signs of myelitis or encephalitis or dermatological complications (involvement of the cornea, zoster generalisatus) have developed as well as in immunocompromised patients an intravenous therapy of virostatic drugs should be applied.

The pain in acute herpes zoster may have a sympathetically maintained component, i.e. some percentage of the pain depends on the efferent sympathetic innervation to the affected skin area. If the zoster is located in the face, the upper or lower extremity diagnostic sympathetic blocks at the stellate ganglion or the lumbosacral chain can be performed particularly in severe cases. In case of a sympathetically maintained pain component a series of several sympathetic blocks may reduce acute pain. The important question whether sympathetic blocks are capable of reducing the risk of chronic PHN is still unresolved.

Prediction of Postherpetic Neuralgia

Despite the similarity of the acute symptoms only a subgroup of zoster patients are at risk for the development of postherpetic neuralgia (PHN). Until now there are no firm indicators the clinician can rely on in setting the prognosis for the patient with acute herpes zoster. However, some interesting associations have started to emerge. In general, patients of older age have a higher risk to be left with PHN [1–3]. In acute herpes zoster there is a positive correlation between elevation of T-lymphocyte CD4/CD8, indicative of the impairment of cell mediated immunity, and duration of pain [4]. Several psychosocial antecedents of the development of chronic pain could be established in herpes zoster, i.e. disease conviction and depression [2]. Furthermore, neurophysiological measures, i.e. elevation of thermal thresholds within the acutely affected skin area, were associated with reports of pain at 3 months duration but the small number of patients precluded assessment of association with chronic established PHN [5]. By analogy with other chronic pain syndromes, i.e. phantom limb pain [6] and chronic back pain [7], many investigators suggested that the acute pain intensity, indicative of the severity of the herpes zoster infection and nerve damage, may predict the development of PHN [2, 3]. In addition, it was shown that patients with a preexisting large fiber polyneuropathy were at higher risk to develop PHN [8].

Prevention of Postherpetic Neuralgia

Recent results from the shingles prevention study shows that vaccination of adults is effective in prevention of PHN [9]. If vaccination of adults is adopted, and shows similar long-term effects, it is possible that there could be a dramatic reduction of PHN in decades to come. Combining this effect with the long-term effect (40+ years) of childhood vaccination against varicella, future generations might be immune to the sometimes devastating results of PHN.

Early antiherpetic therapy, in particular with modern antiviral drugs [10] significantly reduced the development of PHN [11].

The immediate treatment objective during the acute zoster phase is to shorten as much as possible the duration of the pain phase. Although only few controlled studies are available that assessed the effect of acute pain therapy on development of PHN all modern concepts of pain generation suggest that every acute pain input to the nervous system will lead to chronification.

According to the pathophysiological mechanisms several therapeutical interventions to prevent PHN can be hypothesized: (1) an adequate analgesia should be established in the acute phase with e.g. analgesics, anti-depressants or epidural or sympathetic blocks and (2) C-fiber degeneration should be prevented by reducing the inflammatory reaction with e.g. antiviral drugs. So far one controlled study performed so far demonstrated that the incidence of PHN can be reduced by half if 50 mg amitriptyline is administered within the acute zoster phase [12]. Furthermore, gabapentin showed promising results in animal experiments [13].

Taken these data together, a combination of an antiviral agent, analgesics (NSAID or opioids), antidepressants and anticonvulsants and in selected cases sympathetic blocks to treat the sympathetically maintained pain component seems to be appropriate to minimize the risk of pain chronification and development of PHN.

Treatment of Postherpetic Neuralgia

The number of treatment options for PHN has expanded greatly in the last few years [14, 15]. Of particular note are the results of randomized, controlled clinical trials that now confirm the efficacy of anti-depressants, opioids, anticonvulsants and topical analgesics in relieving the symptoms of PHN.

Antidepressants

Tricyclic antidepressants (TCAs) are effective in the treatment of postherpetic pain [16–18]. These compounds are inhibitors of the reuptake of monoaminergic transmitters. They are believed to potentiate the effects of biogenic amines in CNS pain modulating, in particular pain-inhibiting pathways projecting from the brain stem to the spinal cord. In addition, they block voltage dependent Na-channels and alpha adrenergic receptors. However, it may be that the effectiveness of TCAs in neuropathic pain has to do with their broad range of pharmacological actions.

Of the TCAs, amitriptyline is currently the most widely prescribed and best studied compound for the treatment of chronic pain. There is extensive evidence that amitriptyline produces pain relief in PHN [17]. All components of neuropathic pain such as stimulus-independent continuous burning or shooting pain as well stimulus-induced allodynia may be improved. The mean dose required for pain reduction (75–150 mg/day) is usually smaller than doses necessary to achieve anti-depressant effects.

Amitriptyline and other TCAs, however, have significant side-effects. They can produce orthostatic hypotension, largely due to an α-adrenergic blocking action. Due to its histamine receptor blockade, amitriptyline is also a potent sedating drug, which can be a desirable action if patients are having difficulty sleeping. Other significant problems include urinary retention, memory loss and cardiac conduction abnormalities (largely due to the muscarinic anti-cholinergic actions of the drug). Patients, especially the elderly, who are to be treated with this drug should be started at a very low dose (e.g. 10 mg), and built up slowly.

Desipramine and nortriptyline, both of which have predominant norepinephrine reuptake blocking action, appear to be as effective as amitriptyline in PHN [17]. Patients respond to desipramine and nortriptyline at doses comparable to those of amitriptyline but with fewer anti-cholinergic side-effects and significantly less sedation. Still, the side effect profile of the TCAs as a class will continue to represent a significant limitation to their use in the treatment of PHN.

The selective serotonin reuptake inhibitors (SSRI) are an alternative class, but there are as yet no controlled clinical trials with these agents in PHN. In other neuropathic pain states the results with SSRI are disappointing.

There are some newer antidepressants that are neither TCAs nor SSRIs. Venlafaxine and duloxetine block both serotonin and norepinephrine reuptake and have demonstrated efficacy in painful diabetic neuropathy [19, 20].

Based on available data, amitriptyline is still a first-line antidepressant agent in the treatment of PHN. If it is effective but produces intolerable side effects, a cautious trial of nortriptyline or desipramine may be appropriate. Alternatively, a lower dose of amitriptyline may still provide benefit, especially when combined with other types of agents.

Anticonvulsants (Na-Channel Blockers)

Carbamazepine and oxcarbazepine are very effective in trigeminal neuralgia. However, there are no controlled studies in PHN [17]. Newer anticonvulsants like lamotrigine also have some utility in the treatment of peripheral and central neuropathic pain, however, the evidence supporting their use in PHN is currently missing.

Anticonvulsants (Ca-Channel Modulators)

There is a large body of clinical evidence for the efficacy of gabapentin in a variety of neuropathic pain syndromes. Placebo-controlled trials show that gabapentin is effective in PHN [21]. Its relatively benign side effect profile compared to other options have encouraged many physicians to use it frequently for nerve injury pain.

Pregabalin, the successor drug of gabapentin was shown to be efficacious in PHN, DPN and spinal cord injury (until now 7 published studies) [22, 23]. Its mechanism of action has now been solved: a modulating action on the $\alpha 2\delta$-subunit of central Ca-channels located presynaptically at the nociceptive terminal in the dorsal horn spinal cord. Pregabalin has a low potential for drug–drug interactions, and no negative impact on cardiac function. In addition, pregabalin was noted considerably to improve sleep disturbances in neuropathic pain patients. Furthermore, overall mood and other measures of quality of life were positively affected. All these features make it suitable first-line therapy than TCAs or traditional anticonvulsants especially for the elderly, a population very often suffering from several comorbidities that need multiple drug therapies. One advantage over gabapentin is its superior bioavailability which makes it easier to use without the need of long titration periods. Dizziness and somnolence are the most commonly reported adverse events, especially during upward titration to targeted doses.

Opioid Analgesics

Opioids are clearly effective in postoperative, inflammatory and cancer pain. However, the use of opioids for patients with chronic neuropathic pain is controversial, even among experts in the field of pain management, primarily due to a perceived lack of efficacy, and concern about the potential for drug tolerance and addiction.

However, double-blind placebo controlled studies have now demonstrated that acute infusions of morphine or fentanyl give significant pain relief to patients with PHN [24]. Furthermore, recent controlled trials have demonstrated sustained efficacy for several weeks of oral oxycodone [25] and tramadol [26] in PHN. In one study oral morphine was analyzed in a group of PHN patients comparing the effect of antidepressants in the same cohort. Both drugs were similar effective. However, there was no correlation in the response rate between both drugs indicating that different mechanisms are active in these PHN patients [27].

All data on opioid use in chronic nonmalignant pain collected so far are insufficient to address the long-term efficacy of opioids and the development of adverse effects that might only arise during long-term use, e.g. their effect on the immune system. However, many patients with pain due to central and peripheral nerve injury can be successfully and safely treated on a chronic basis with stable doses of strong opioids without signs of tolerance. The use of opioids requires caution in patients with a history of chemical dependence or pulmonary disease. We recommend using long-acting opioid analgesics (e.g. sustained release morphine preparation) when alternative approaches to treatment have failed. An opioid trial should be tested before invasive therapies are instituted. Furthermore, a trial of opioids should not be delayed to a 'last resort' status. Prophylactic treatment of common side effects notably nausea or constipation is necessary and improves patients' compliance.

NMDA-Receptor Antagonists

These drugs block excitatory glutamate receptors in the CNS that thought to be responsible for the increased central excitability (central sensitization) following noxious stimuli. Clinically available substances with NMDA receptor blocking properties include ketamine, dextromethorphane, memantine and amantadine. Typical side effects include sedation, nausea, disagreeable psychological disturbances or even frank hallucinations. Dextromethorphan, memantine and amantadine have fewer side affects.

Studies of small cohorts have generally confirmed the analgesic effects of ketamine in patients suffering from PHN [28]. However, studies with oral NMDA-antagonists formulations (e.g. dextromethorphane) showed positive results in painful diabetic neuropathy but the drug was without beneficial effect in PHN [29, 30].

Topical Medications

Topical Capsaicin: Capsaicin is an agonist of the vanilloid receptor which is present on the sensitive terminals of primary nociceptive afferents. On initial application it has an excitatory action and produces burning pain and hyperalgesia, but with repeated or prolonged application it inactivates the receptive terminals of nociceptors. Therefore, this approach is reasonable for those patients whose pain is maintained by anatomically intact sensitized nociceptors.

Capsaicin extracts are available in a 0.025 and 0.075% preparation. The 0.025 and 0.075% preparations have been reported to reduce the pain of PHN [31, 32]. Capsaicin preparations often produce intolerable burning so that many patients discontinue their use.

Topical Lidocaine: A second topical medication for neuropathic pain are local anesthetics. Local anesthetics block voltage-dependent Na-channels. Although the site of action of membrane-stabilizing drugs for relief of pain has not been proven in patients, in vitro studies have shown that ectopic impulses generated by damaged primary afferent nociceptors are abolished by concentrations of local anesthetics much lower than that required for blocking normal axonal conduction.

Controlled studies report pain relief with topically applied special formulations of local anesthetic. Lidocaine patches (5%) were evaluated in several controlled studies [33–35]. Pain relief was statistically significant compared with the control group between 4 and 12 h following application of the patch. Blood levels of lidocaine were at least an order of magnitude below those required for an anti-arrhythmic effect, and therefore there were only minor adverse effects associated with application of the patch itself. Lidocaine patch therapy is a safe and well-tolerated supplemental modality for PHN pain relief.

Intrathecally Administered Drugs

Intrathecal administration of lidocaine and methyl prednisolone combined appear to be associated with remarkable benefit in PHN patients [36]. However, the therapy has potentially dangerous side effects and the trial has not yet been replicated. Therefore, it is suggested to wait for further high-quality controlled trials for this therapy before definite recommendations can be made [15].

Stimulation Techniques

Transcutaneous electrical nerve stimulation (TENS) may be effective in some cases and has minimal side effects. It should be avoided to place the electrodes within skin areas with allodynia since pain may be exaggerated. Alternatively the electrodes may be fixed at adjacent dermatomes or even contralaterally.

Invasive stimulation techniques, epidural spinal cord stimulation and deep brain stimulation (sensory thalamus, motor cortex), have been reported to be effective in selected cases of PHN.

Treatment Guideline

In summary, adult vaccination seem to be effective for prevention of shingles and PHN. In acute herpes zoster early antiviral therapy is recommended and immediate pain treatment should be initiated. The following treatment algorithm for PHN (fig. 1) is based on the results of available controlled trials in PHN, several recent meta-analyses of therapy of neuropathic pain and clinical experience (table 1) [14, 15, 17, 37]. The medical management of PHN consists of four main classes of oral medication (serotonin/norepinephrine reuptake blockers, Na-channel-anticonvulsants, Ca-channel-anticonvulsants, opioids) and several categories of topical medications for patients with cutaneous allodynia and hyperalgesia (capsaicin and local anesthetics). However, it should be noted that so far no controlled trials exists for carbamezepine, oxcarbazepine, lamotrigine, duloxetine and venlafaxine and most opioids in PHN (table 1).

Since more than one mechanism of PHN is at work in most patients, a combination of two or more analgesic agents to cover multiple types of mechanisms will generally produce greater pain relief and fewer side effects. Therefore, early combinations of two or three compounds out of different classes may be more appropriate for some patients instead of a stepwise proceeding with a successive monotherapy. This is indicated in the circles in figure 1. Indeed, in a recent controlled four-period crossover trial gabapentin and morphine combined achieved better analgesia at lower doses of each drug than either as a single agent with constipation, sedation and dry mouth as the most frequent adverse effects [38].

In particular cases intrathecal administration of lidocaine and methyl prednisolone combined or invasive stimulation techniques like epidural spinal cord

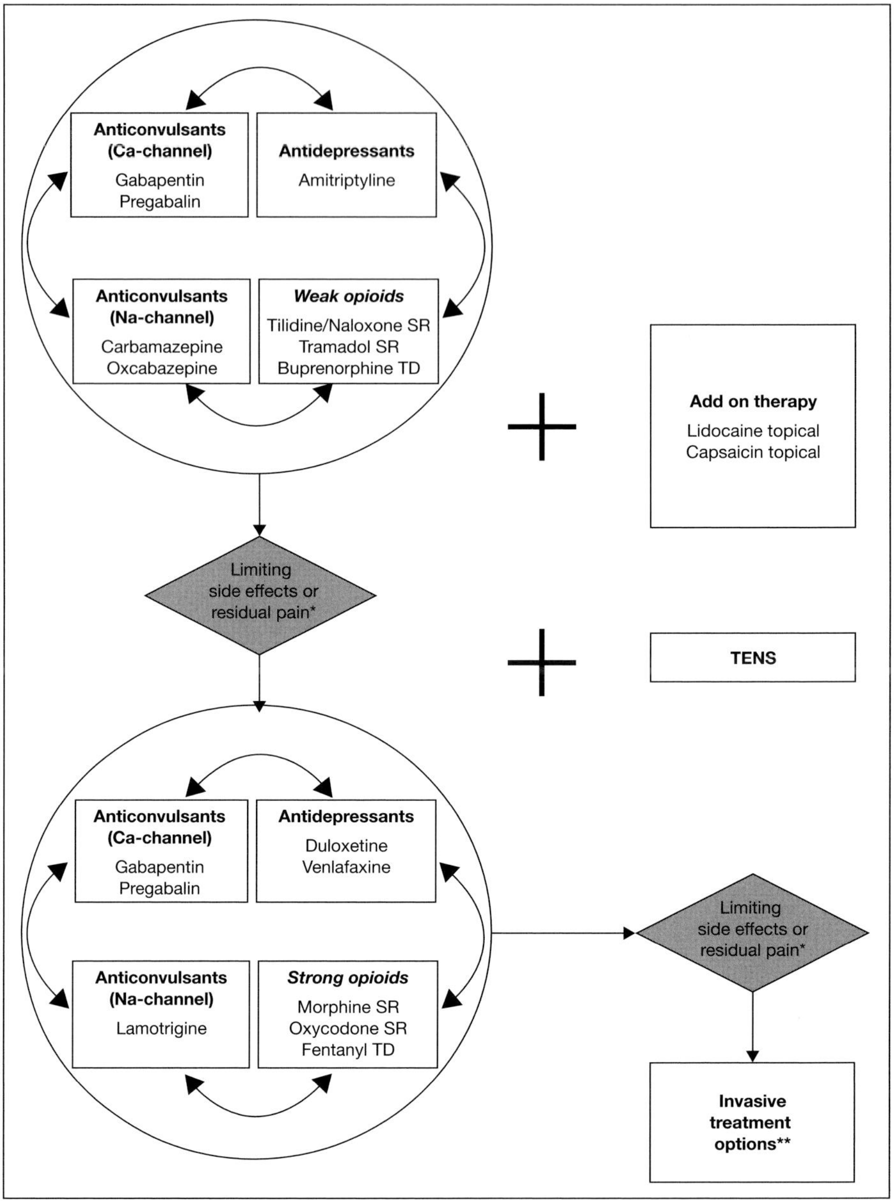

Fig. 1. Algorithm for the treatment of neuropathic pain. An early combination of two or more agents to cover multiple types of mechanisms will generally produce greater pain relief and fewer side effects. This is illustrated in the circles. *Pain level significant and persistent for at least 2–4 weeks. **Spinal cord stimulation, deep brain stimulation or motor cortex stimulation. TENS = transcutaneous electrical nerve stimulation.

Table 1. Pharmacological therapy of postherpetic neuralgia (doses for adults)

Compound	Efficacious dose (maximal dose) [mg/day]	Dose interval	Evidence
Anti-depressants			
Amitriptyline	50–75 (150)	0–0–1	⇑⇑
Anti-convulsants (Ca-channel)			
Gabapentin	1,200–2,400 (3,600)	1–1–1	⇑⇑
Pregabalin	150 (600)	1–0–1	⇑⇑
Long-acting opioids			
Tramadol SR	Titration (600)	1–(1)–1	⇑
Morphine SR	Titration no	1–(1)–1	⇑
Oxycodone	Titration no	1–(1)–1	⇑
Topical therapy			
Capsaicin cream	–	4 × die	⇑
Lidocain-patch	–	(3 patches/die) 1 × die	⇑⇑

Levels of evidence.
⇑ = at least 1 RCT; ⇑⇑ = Several RCT or metaanalyses.

stimulation may be indicated. Transcutaneous electrical nerve stimulation may be effective in some cases and has minimal side effects.

However, beyond these treatment approaches the importance of the biopsychosocial model of chronic pain should be considered by additional management of psychological and social aspects [15, 39].

The treatment of neuropathic pain is currently still unsatisfactory. The hope is that in the future novel drugs will be developed that address specifically the relevant combination of mechanisms in one particular patient leading to an optimal individual polypragmatic therapy [40].

References

1 Watson CP, Watt VR, Chipman M, Birkett N, Evans RJ: The prognosis with postherpetic neuralgia. Pain 1991;46:195–199.
2 Dworkin RH, Hartstein G, Rosner HL, Walther RR, Sweeney EW, Brand L: A high-risk method for studying psychosocial antecedents of chronic pain: the prospective investigation of herpes zoster. J Abnorm Psychol 1992;101:200–205.
3 Bruxelle J: Prospective epidemiologic study of painful and neurologic sequelae induced by herpes zoster in patients treated early with oral acyclovir. Neurology 1995;45(suppl 8):S78–S79.
4 Higa K, Noda B, Manabe H, Sato S, Dan K: T-lymphocyte subsets in otherwise healthy patients with herpes zoster and relationships to the duration of acute herpetic pain. Pain 1992;51:111–118.

5 Nurmikko T, Bowsher D: Somatosensory findings in postherpetic neuralgia. J Neurol Neurosurg Psychiatry 1990;53:135–141.
6 Jensen TS, Krebs B, Nielsen J, Rasmussen P: Immediate and long-term phantom limb pain in amputees: incidence, clinical characteristics and relationship to pre-amputation limb pain. Pain 1985;21:267–278.
7 Gatchel RJ, Polatin PB, Kinney RK: Predicting outcome of chronic back pain using clinical predictors of psychopathology: a prospective analysis. Health Psychol 1995;14:415–420.
8 Baron R, Haendler G, Schulte H: Afferent large fiber polyneuropathy predicts the development of postherpetic neuralgia. Pain 1997;73:231–238.
9 Oxman MN, Levin MJ, Johnson GR, Schmader KE, Straus SE, Gelb LD, Arbeit RD, Simberkoff MS, Gershon AA, Davis LE, Weinberg A, Boardman KD, Williams HM, Zhang JH, Peduzzi PN, Beisel CE, Morrison VA, Guatelli JC, Brooks PA, Kauffman CA, Pachucki CT, Neuzil KM, Betts RF, Wright PF, Griffin MR, Brunell P, Soto NE, Marques AR, Keay SK, Goodman RP, Cotton DJ, Gnann JW, Jr., Loutit J, Holodniy M, Keitel WA, Crawford GE, Yeh SS, Lobo Z, Toney JF, Greenberg RN, Keller PM, Harbecke R, Hayward AR, Irwin MR, Kyriakides TC, Chan CY, Chan IS, Wang WW, Annunziato PW, Silber JL: A vaccine to prevent herpes zoster and postherpetic neuralgia in older adults. N Engl J Med 2005;352:2271–2284.
10 Wassilew S: Brivudin compared with famciclovir in the treatment of herpes zoster: effects in acute disease and chronic pain in immunocompetent patients. A randomized, double-blind, multinational study. J Eur Acad Dermatol Venereol 2005;19:47–55.
11 Wood MJ, Kay R, Dworkin RH, Soong SJ, Whitley RJ: Oral acyclovir therapy accelerates pain resolution in patients with herpes zoster: a meta-analysis of placebo-controlled trials. Clin Infect Dis 1996;22:341–347.
12 Bowsher D: The effects of pre-emptive treatment of postherpetic neuralgia with amitriptyline: a randomized, double-blind, placebo-controlled trial. J Pain Symptom Manage 1997;13:327–331.
13 Kuraishi Y, Takasaki I, Nojima H, Shiraki K, Takahata H: Effects of the suppression of acute herpetic pain by gabapentin and amitriptyline on the incidence of delayed postherpetic pain in mice. Life Sci 2004;74:2619–2626.
14 Dworkin RH, Schmader KE: Treatment and prevention of postherpetic neuralgia. Clin Infect Dis 2003;36:877–882.
15 Hempenstall K, Nurmikko TJ, Johnson RW, A'Hern RP, Rice AS: Analgesic therapy in postherpetic neuralgia: a quantitative systematic review. PLoS Med 2005;2:e164.
16 McQuay HJ, Tramer M, Nye BA, Carroll D, Wiffen PJ, Moore RA: A systematic review of antidepressants in neuropathic pain. Pain 1996;68:217–227.
17 Sindrup SH, Jensen TS: Efficacy of pharmacological treatments of neuropathic pain: an update and effect related to mechanism of drug action. Pain 1999;83:389–400.
18 Collins SL, Moore RA, McQuay HJ, Wiffen P: Antidepressants and anticonvulsants for diabetic neuropathy and postherpetic neuralgia: a quantitative systematic review. J Pain Symptom Manage 2000;20:449–458.
19 Goldstein DJ, Lu Y, Detke MJ, Lee TC, Iyengar S: Duloxetine vs. placebo in patients with painful diabetic neuropathy. Pain 2005;116:109–118.
20 Rowbotham MC, Goli V, Kunz NR, Lei D: Venlafaxine extended release in the treatment of painful diabetic neuropathy: a double-blind, placebo-controlled study. Pain 2004;110:697–706.
21 Rowbotham M, Harden N, Stacey B, Bernstein P, Magnus-Miller L: Gabapentin for the treatment of postherpetic neuralgia: a randomized controlled trial. JAMA 1998;280:1837–1842.
22 Dworkin RH, Corbin AE, Young JP: Pregabalin for the treatment of postherpetic neuralgia: a randomized, placebo-controlled trial. Neurology 2003;22:1274–1283.
23 Sabatowski R, Galvez R, Cherry DA, Jacquot F, Vincent E, Maisonobe P, Versavel M: Pregabalin reduces pain and improves sleep and mood disturbances in patients with post-herpetic neuralgia: results of a randomised, placebo-controlled clinical trial. Pain 2004;109:26–35.
24 Rowbotham MC, Reisner-Keller LA, Fields HL: Both intravenous lidocaine and morphine reduce the pain of postherpetic neuralgia. Neurology 1991;41:1024–1028.
25 Watson CP, Babul N: Efficacy of oxycodone in neuropathic pain: a randomized trial in postherpetic neuralgia [In Process Citation]. Neurology 1998;50:1837–1841.

26 Göbel H, Stadler T: Treatment of post-herpes zoster pain with tramadol. Results of an open pilot study versus clomipramine with or without levomepromazine. Drugs 1997;53(suppl 2):34–39.
27 Raja SN, Haythornthwaite JA, Pappagallo M, Clark MR, Travison TG, Sabeen S, Royall RM, Max MB: Opioids versus antidepressants in postherpetic neuralgia: a randomized, placebo-controlled trial. Neurology 2002;59:1015–1021.
28 Eide K, Stubhaug A, Oye I, Breivik H: Continuous subcutaneous administration of the N-methyl-D-aspartic acid (NMDA) receptor antagonist ketamine in the treatment of post-herpetic neuralgia. Pain 1995;61:221–228.
29 Nelson KA, Park KM, Robinovitz E, Tsigos C, Max MB: High-dose oral dextromethorphan versus placebo in painful diabetic neuropathy and postherpetic neuralgia. Neurology 1997;48:1212–1218.
30 Sang CN, Booher S, Gilron I, Parada S, Max MB: Dextromethorphan and memantine in painful diabetic neuropathy and postherpetic neuralgia: efficacy and dose-response trials. Anesthesiology 2002;96:1053–1061.
31 Watson CP, Tyler KL, Bickers DR, Millikan LE, Smith S, Coleman E: A randomized vehicle-controlled trial of topical capsaicin in the treatment of postherpetic neuralgia. Clin Ther 1993;15: 510–526.
32 Bernstein JE, Korman NJ, Bickers DR, Dahl MV, Millikan LE: Topical capsaicin treatment of chronic postherpetic neuralgia. J Am Acad Dermatol 1989;21(pt 1):265–270.
33 Galer BS, Rowbotham MC, Perander J, Friedman E: Topical lidocaine patch relieves postherpetic neuralgia more effectively than a vehicle topical patch: results of an enriched enrollment study. Pain 1999;80:533–538.
34 Meier T, Wasner G, Faust M, Kuntzer T, Ochsner F, Hueppe M, Bogousslavsky J, Baron R: Efficacy of lidocaine patch 5% in the treatment of focal peripheral neuropathic pain syndromes: a randomized, double-blind, placebo-controlled study. Pain 2003;106:151–158.
35 Rowbotham MC, Davies PS, Verkempinck C, Galer BS: Lidocaine patch: double-blind controlled study of a new treatment method for post-herpetic neuralgia. Pain 1996;65:39–44.
36 Kotani N, Kushikata T, Hashimoto H, Kimura F, Muraoka M, Yodono M, Asai M, Matsuki A: Intrathecal methylprednisolone for intractable postherpetic neuralgia. N Engl J Med 2000;343: 1514–1519.
37 Sindrup SH, Jensen TS: Pharmacologic treatment of pain in polyneuropathy. Neurology 2000;55: 915–920.
38 Gilron I, Bailey JM, Tu D, Holden RR, Weaver DF, Houlden RL: Morphine, gabapentin, or their combination for neuropathic pain. N Engl J Med 2005;352:1324–1334.
39 Haythornthwaite JA, Benrud-Larson LM: Psychological aspects of neuropathic pain. Clin J Pain 2000;16(suppl):S101–S105.
40 Fields HL, Rowbotham M, Baron R: Postherpetic neuralgia: irritable nociceptors and deafferentation. Neurobiol Dis 1998;5:209–227.

Prof. Dr. Med. Ralf Baron
Klinik für Neurologie, UK-SH, Campus Kiel
Sektion Neurologische Schmerzforschung und Therapie
Schittenhelmstr. 10
DE–24105 Kiel (Germany)
Tel. +49 431 597 8505, Fax +49 431 597 8530, E-Mail r.baron@neurologie.uni-kiel.de

Gross G, Doerr HW (eds): Herpes Zoster.
Monogr Virol. Basel, Karger, 2006, vol 26, pp 154–163

Epidemiology of Herpes Zoster: What has Changed?

Regina Allwinn, Sigune Buxbaum, Hans Wilhelm Doerr

Institute for Medical Virology, University Hospital of Frankfurt, Frankfurt am Main, Germany

Herpes zoster (shingles) is an inflammatory neurodermatologic disease, usually localized on a skin segment of the body which is innervated by a sensory nerve. More than 100 years ago, an association between Varicella and herpes zoster has been suggested (see preface of the book). Shingles is a secondary disease to passed Varicella virus (VZV) infection, which latently persists lifelong in the spinal ganglia of the host. The presumably proviral latency is switched to productive infection by several trigger factors resulting in shingles ('girdle rose'). The most important trigger factor is a waning cell-mediated immunity to VZV along a big time interval after primary infection during childhood. Thus, the majority of patients are elderly or those who suffer from immunocompromising diseases. Nevertheless, many case reports remind that herpes zoster occurs also in immunocompetent adolescents and even in children. Numerous clinical observations have elucidated the epidemiology of herpes zoster. Figure 1 displays the viral circulation through the population.

Seroepidemiology

Similar to many other viral infections the spread of VZV can be estimated by serum surveys, which means determination of IgG antibody prevalence in age grouped population samples. IgG antibodies are formed after varicella or subclinic VZV infection and persist lifelong in stable titers. Numerous prospective studies or retrospective evaluations have shown that high prevalences are reached already in childhood [1–3]. During the course of life, fluctuating antibody titer is no marker for a subsequent development of herpes zoster. Figure 2 shows serum surveys on different seroconversion titers of VZV-specific IgG antibodies and

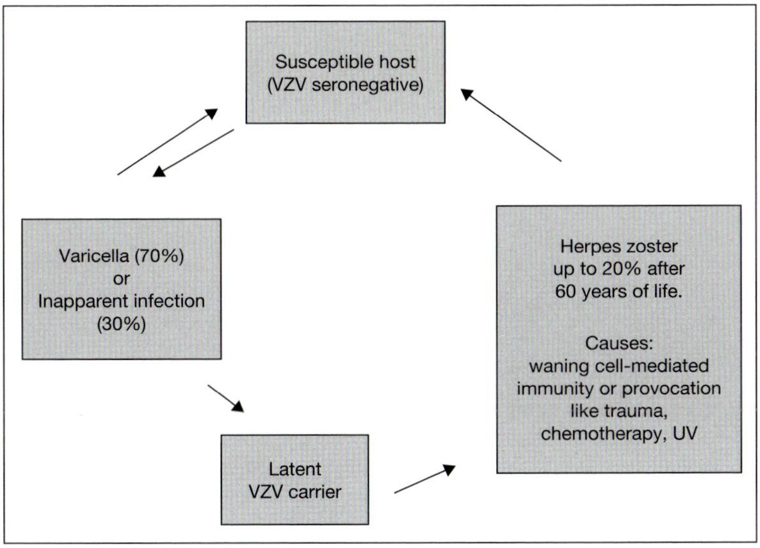

Fig. 1. Natural spread of VZV infection.

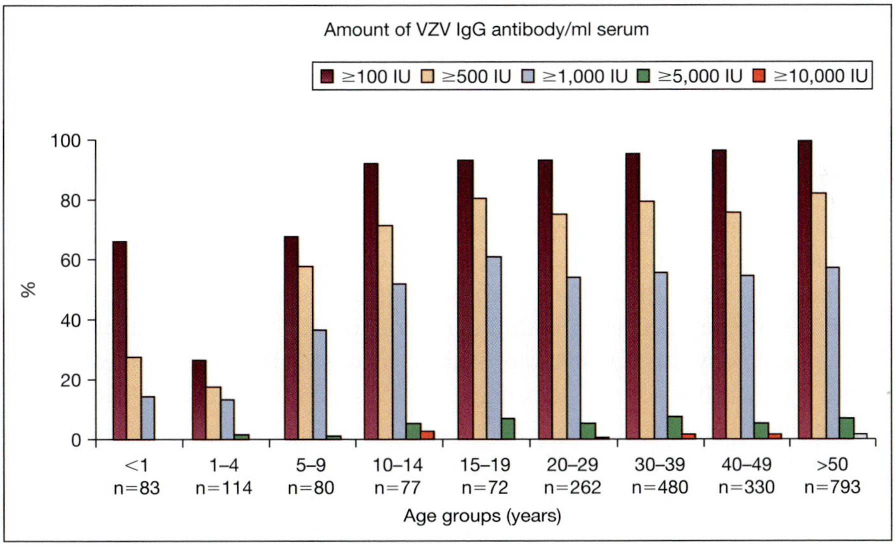

Fig. 2. VZV-specific IgG seroprevalence in difference age groups. University Hospital Frankfurt/Main (from reference [1]).

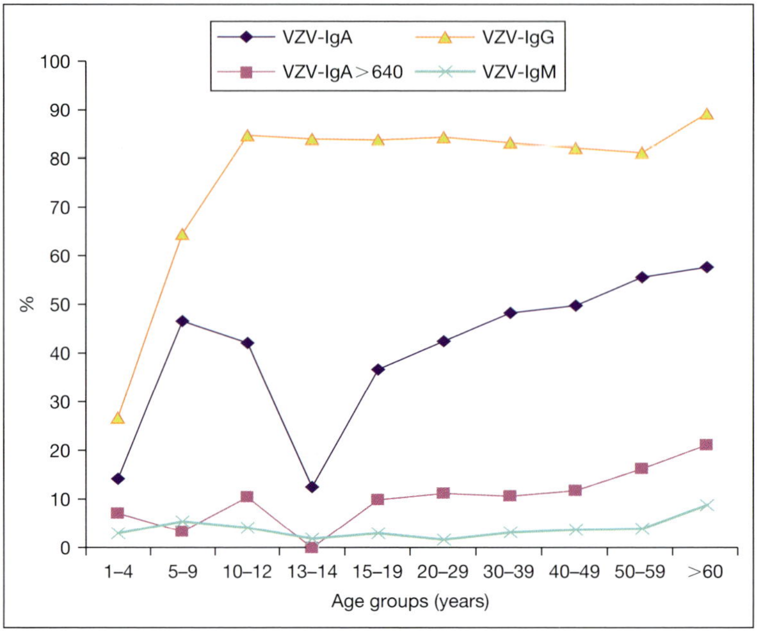

Fig. 3. Varicella–zoster antibody seroprevalence (IgA, IgM and IgG), 1999–2004 (n = 5,457 serum samples). University Hospital Frankfurt/Main.

reveals that titer distribution is similar in all age groups [1]. The rapid spread of VZV has not been reduced by improvement of hygiene and social-economic status. For example, in Germany the rise of antibody prevalence throughout life remained constant for 25 years, as compared to a study from 1973 to 1974 [1, 4]. HIV carriers has statistically significant higher prevalences of VZV-specific antibodies, further increasing with progression to AIDS [5]. IgM antibodies and high IgA antibody titers are markers of an active or reactivated VZV infection [6]. Figure 3 shows that – at low percentage – VZV specific IgM and IgA antibodies are present in the serum survey of all age groups and rising in the elderly as expected. This corresponds to the classical scheme given by Hope-Simpson (fig. 4). In contrast to recurrent herpes simplex, VZV reactivations occur throughout life, but they remain sub-clinical for a long time. Only when cell-mediated immunity is reduced, the stimulus of reactivation is high enough to make a breakthrough resulting in a zoster disease [7]. Thus, by serological means, herpes zoster epidemiology can be only approximated. The exact diagnosis of shingles requires clinical evaluation in addition to laboratory markers.

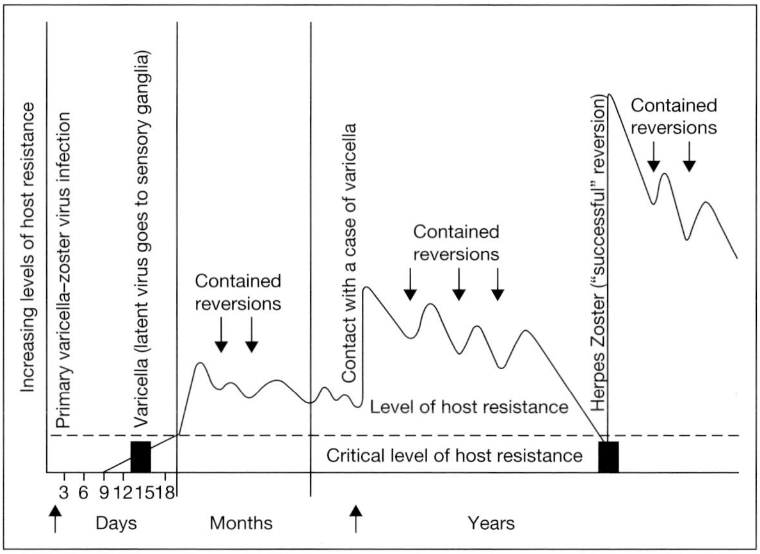

Fig. 4. Immune response after infection VZV (from reference [12]).

Clinical Epidemiology

Before the introduction of VZV vaccination, only a few *population-based* studies of the natural history and epidemiology of herpes zoster have been conducted [8]. The two classic investigations have been performed in UK and USA by Hope-Simpson [9, 10] and by Ragozzino et al. [11] who retrospectively analyzed records of the time periods 1947–1962 and 1945–1959, respectively. Beginning with the first year of life, the cumulative incidence of herpes zoster rises slowly in an exponential way (fig. 5). After the 60th year of life, at least 20% of the population has once passed an episode of herpes zoster; only a few of them have presented a second episode. Even in infants some cases are observed, probably as a consequence of prenatal or early childhood VZV infection.

Now, 30 years later, the herpes zoster epidemiology has obviously changed. Donahue et al. [12] investigated incident and recurrent herpes zoster from 1990 to 1992 in a health maintenance organization and found significant higher incidences of shingles. A 'new' epidemiology of shingles was also ruled out in Central Europe by the 'German Zoster Study Group' [13] which *prospectively* recruited 2,063 patients suffering from shingles in different dermatomes between September 1994 and March 1995. Nearly 20% of patients were younger than 30 years of age! Since 1990, it was observed, that children are more afflicted by herpes zoster, going parallel with a broader use of corticoids

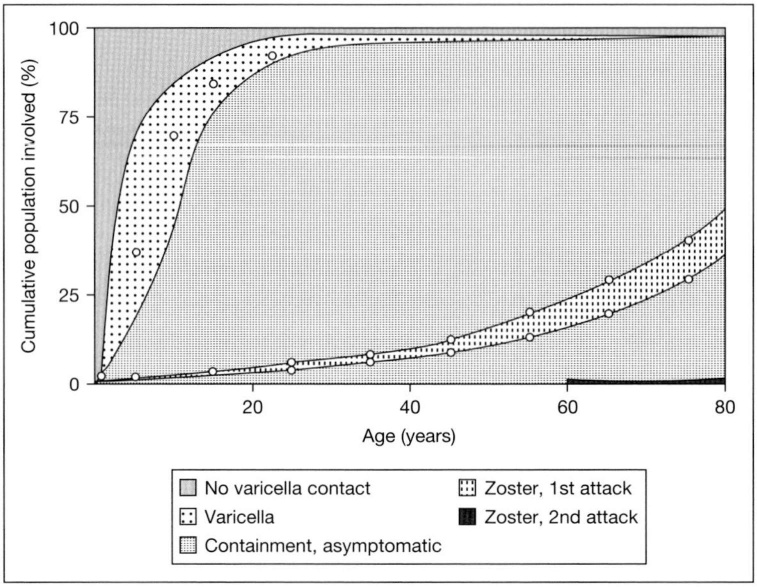

Fig. 5. Age-related activity of VZV in the Anglo-American population, 1947–1962 (from reference [11]).

[14]. So, today it is not justified to consider shingles only as a geriatric problem (fig. 6). The proportion of male and female zoster patients was found equal. The larger number of women in the highest age group (>70 years) reflects the age distribution of the German population.

While age distribution of herpes zoster patients has moved to the younger, a remarkable stability was found over the time span of about 40 years between the three studies of Hope-Simpson [9, 10], Ragozzino et al. [11] and Meister et al. [13, 16] regarding the localization of shingles (fig. 7). Thoracic herpes zoster predominates over cranial, cervical, lumbar and sacral manifestations. However, the cranial, lumbar and in particular, the sacral manifestations were registered relatively more often in the previous study [13, 16]. This finding is age-dependent (fig. 8). The thoracic shingles decrease, while the other localizations increase during age of life.

The frequency of severe courses in herpes zoster patients is of special clinical and epidemiological interest. Between 1991 and 2000, among 373 cases of herpes zoster per 100,000 man-years, the hospitalization rate in England and Wales was 4.4 cases (similar to Varicella), 8% having underlying immunosuppressive conditions [15].

The most severe complication of shingles is the development of postzosteric neuralgia (PZN). The complicated courses of shingles observed in the

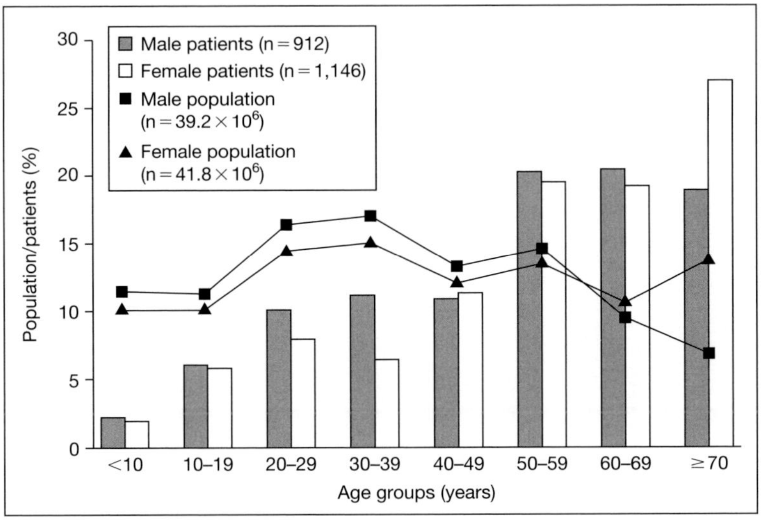

Fig. 6. Age and sex distribution among patients with herpes zoster compared with the data from the German general population in 1993 (from reference [13]).

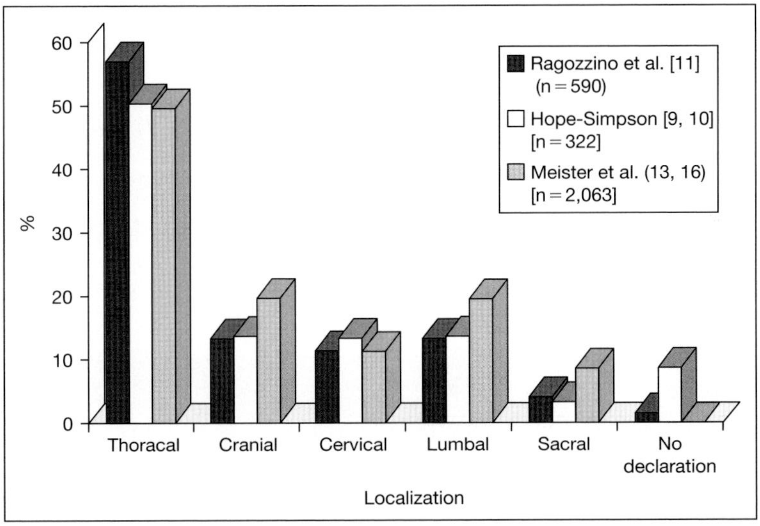

Fig. 7. Localization of herpes zoster recorded in three different studies.

German study are compiled in table 1. To summarize, 320 among 2,063 patients (15.5%) required more intensive medical treatment.

Older and female patients had been more afflicted by PZN. Furthermore, the number, as well as the hemorrhages of skin lesions with cranial or sacral

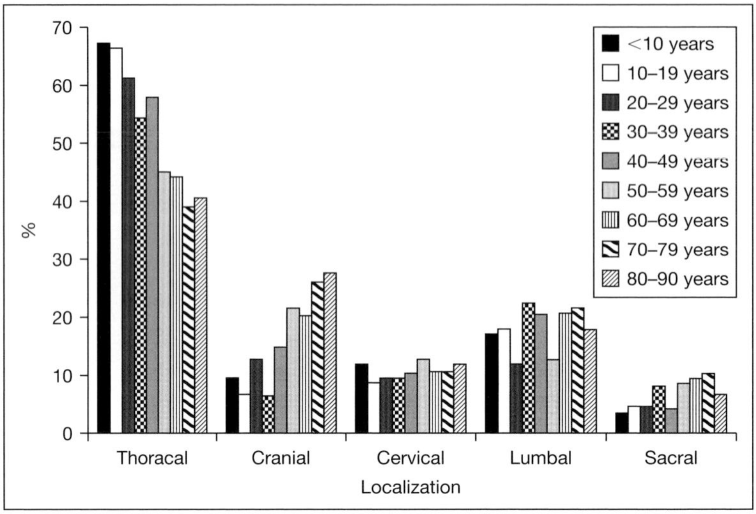

Fig. 8. Age-dependency of zoster localization (from reference [13]).

Table 1. Complicated courses of herpes zoster (modified from reference [16])

Complication	Number of cases
Visceral involvement	16
Zoster ophthalmicus	55
Zoster oticus	18
Paresis	41
Extension to further dermatomes	24
Postzosteric neuralgia (PZN)	135
Other (not further specified)	31
Total complications	*320*

localizations of shingles were recognized as risk factors. This could be compiled to a prognostic score for PZN [16].

Although population-based epidemiology has changed, herpes zoster is a mandatory indication to look for an immunocompromising status, if the patient is younger than 35 years. Shingles are a marker of progredient immunodeficiency in HIV carriers. Further immunocompromising conditions leading to shingles are leukemia and Hodgkin's disease.

A similar high risk factor is therapeutic immunosuppression after organ and in particular after bone marrow transplantation (see chapter: Herpes Zoster in the Immunocompromised Host, pp 93–106). Thus, these people are recommended to be vaccinated or screened on natural immunity before immunsuppression (see chapter: Live Attenuated Varicella Vaccine, pp 164–169).

Change of Epidemiology after General VZV Vaccination

After the successful vaccination of risk groups against VZV infection, general immunization has been introduced into the USA in 1995. By the use of consequent vaccination programs in early childhood, the occurrence of varicella could be dramatically reduced (>90%) and the mortality decreased by two third [17, 18]. In Germany, vaccination against VZV is generally recommended since August 2004. In this way, the threshold of 90% population immunization which is necessary for herd immunity can be reached easier. The vaccine is an attenuated VZV strain (Oka strain) imitating the natural infection and thus inducing immunity in a similar way. After immunization the vaccine virus becomes latent. Later reactivation has been described in immunocompromised and also healthy vaccinees. This usually appears as a mild course of herpes zoster [19]. The persistence of vaccine-induced immunity does not last as long as natural immunity. A growing number of 'breakthrough Varicella' among immunized children was recognized [20–22].

Reactivation of vaccine virus or breakthrough of wild virus infection should not limit the decrease of natural herpes zoster. However, if wild VZV circulation is interrupted, the sub-clinical booster of immunity against natural herpes zoster is reduced. Before varicella eradication, the huge pool of latent wild VZV represents a threat, namely that herpes zoster could become more frequent [7]. Paradoxically, a good VZV herd immunity of 90% could increase herpes zoster incidences, which has been calculated with mathematical models [23].

Using the random-digit-dial Behavioral Risk Factor Surveillance System, Yih et al. [24] monitored the incidence of varicella and herpes zoster in Massachusetts between 1998 and 2003. They confirmed that due to general vaccination the incidence of varicella has decreased as expected, while the incidence of herpes zoster has significantly increased from 2.77/1,000 to 5.25/1,000. Thus, until VZV is eradicated, a booster vaccination of the elderly might be recommended. A second vaccination with VZV decreases the herpes zoster incidence, as recently shown in a large study on prevention of shingles and postherpetic neuralgia [19]. At the moment, it cannot be decided whether this is generally needed. More epidemiologic data are necessary for better effective and reliable mathematical modeling [25].

References

1 Buxbaum S, Doerr HW, Allwinn R: Epidemiological analysis of immunity against vaccine-preventable diseases: rubella, measles, mumps and chickenpox. Dtsch Med Wochenschr 2001;126:1289–1293.

2 Mossong J, Putz L, Schneider F: Seroprevalence and force of infection of varicella-zoster virus in Luxembourg. Epidemiol Infect 2004;132:1121–1127.

3 Vyse, AJ, Gay NJ, Hesketh LM, Morgan-Capner P, Miller E: Seroprevalence of antibody to varicella zoster virus in England and Wales in children and young adults. Epidemiol Infect 2004;132: 1129–1134.

4 Doerr HW, Lehmair H, Schmitz H, Kampa D, Luthardt TH: Simple mathematical deductions in the seroepidemiology of viral infections. I. Herpesvirus group (Herpesvirus hominis; Varicella-Zoster-Virus, Cytomegalovirus, Epstein-Barr-Virus). Zentrabl Bakteriol [Orig A] 1977;238: 149–164.

5 Enzensberger R, Braun W, July Ch, Helm EB, Doerr HW: Prevalence of antibodies to human herpesvirus and hepatitis B virus in patients at different stages of human immunodeficiency virus (HIV). Infection 1991;19:140–145.

6 Doerr HW, Rentschler M, Scheifler G: Serologic detection of active infections with human herpes viruses (CMV, EBV, HSV, VZV): diagnostic potential of IgA class and IgG subclass-specific antibodies. Infection 1987;15:93–98.

7 Hambleton S, Gershon AA: Preventing varicella-zoster disease. Clin Microbiol Rev 2005;18:70–80.

8 Weller TH: Varicella and herpes zoster. Changing concepts of the natural history, control, and importance of a not-so-benign virus (First of Two Parts). N Engl J Med 1983;309:1362–1368.

9 Hope-Simpson RE: Postherpetic neuralgia. J R Coll Gen Pract 1975;25:571–575.

10 Hope-Simpson: The nature of herpes zoster. A long-term study and a new hypothesis. Proc R Soc Med 1965;58:9–20.

11 Ragozzino MW, Melton LJ 3rd, Kurland LT, Chu CP, Perry HO: Population-based study of herpes zoster and its sequelae. Medicine (Baltimore) 1982;61:310–316.

12 Donahue JG, Choo PW, Manson JE, Platt R: The incidence of herpes zoster. Arch Intern Med 1995;155:1605–1609.

13 Meister W, Neiss A, Gross G, Doerr HW, Höbel W, Malin JP, von Essen J, Reimann BY, Witke C, Wutzler P: Demography, symptomatology, and course of disease in ambulatory zoster patients. A physican-based survey in Germany. Intervirology 1998;41:272–277.

14 Uebe B, Sauerbrei A, Burdach S, Horneff G: Herpes zoster by reactivated vaccine varicella zoster virus in a healthy child. Eur J Pediatr 2002;161:442–444.

15 Brisson M, Edmunds WJ: Epidemiology of varicella-zoster virus in England and Wales. J Med Virol 2003;70(suppl 1):9–14.

16 Meister W, Neiß A, Gross G, Doerr HW, Höbel W, Malin JP, von Essen J, Reimann BY, Witke C, Wutzler P: A prognostic score for postherpetic neuralgia in ambulatory patients. Infection 1998;26:6.

17 Nguyen HQ, Jumaan AO, Seward JF: Decline in mortality due to varicella after implementation of varicella vaccination in the Unites States. N Engl J Med 2005;352:450–458.

18 Vázques M, Shapiro ED: Varicella vaccine and infection with varicella-zoster virus. N Engl J Med 2005;352:439–440.

19 Oxmann MN, Levin MJ, Johnson GR, Schmader KE, Straus SE, Gelb LD, Arbeit RD, Simberkoff MS, Gershon AA, Davis LE, Weinberg A, Boardman KD, Williams HM, Zhang JH, Peduzzi PN, Beisel CE, Morrison VA, Guatelli JC, Brooks PA, Kauffman CA, Pachucki CT, Neuzil KM, Betts RF, Wright PF, Griffin MR, Brunell P, Soto NE, Marques AR, Keay SK, Goodmann RP, Cotton DJ, Gnann JW Jr, Loutit J, Holodniy M, Keitel WA, Crawford GE, Yeah SS, Lobo Z, Toney JF, Greenberg RN, Keller PM, Harbecke R, Hayward AR, Irwin MR, Kyriakides TC, Chan CY, Chan IS Wang WW, Annunziato PW, Silber JL, Shingles Prevention Group: A vaccine to prevent herpes zoster and postherpetic neuralgia in older adults. N Engl J Med 2005;352:2271–2284.

20 Grose C: Varicella vaccination of children in the United States: assessment after the first decade 1995–2005. J Clin Virol 2005;33:89–95; discussion 96–98.

21 Haddad MB, Hill MB, Pavia AT, Green CE, Jumaan AO, De AK, Rolfs RT: Vaccine effectiveness during a varicella outbreak among school children: Utah, 2002–2003. Pediatrics 2005;115:1488–1493.
22 Ludwig B, Kraus FB, Allwinn R, Keim S, Doerr HW, Buxbaum S: Loss of VZV-antibodies in a health care worker – how to deal with the wane of immunity after vaccination? Infection, submitted.
23 Brisson M, Edmunds WJ, Gay NJ: Varicella vaccination: impact of vaccine efficacy on the epidemiology of VZV. J Med Virol 2003;70(suppl 1):31–37.
24 Yih WK, Brooks DR, Lett SM, Jumaan AO, Zhang Z, Clements KM, Seward JF: The incidence of varicella and herpes zoster in Massachusetts as measured by the Behavioral Risk Factor Surveillance System (BRFSS) during a period of increasing varicella vaccine coverage, 1998–2003. BMC Public Health 2005;5:68.
25 Wagenpfeil S, Neiss A, Wutzler P: Effects of varicella vaccination on herpes zoster incidence. Clin Microbiol Infect 2004;10:954–960.

Professor H.W. Doerr, M.D.
Institute for Medical Virology, University Hospital of Frankfurt
Paul-Ehrlich-Strasse 40
DE–60596 Frankfurt/Main (Germany)
Tel. +49 69 63015219, Fax +49 69 63016244, E-Mail H.W.Doerr@em.uni-frankfurt.de

Gross G, Doerr HW (eds): Herpes Zoster.
Monogr Virol. Basel, Karger, 2006, vol 26, pp 164–169

Live Attenuated Varicella Vaccine

Anne A. Gershon

Department of Pediatrics, Columbia University College of Physicians and Surgeons, New York, N.Y., USA

The Virus and Its Diseases

Varicella–zoster virus (VZV) is the etiologic agent of varicella and zoster. Varicella is the primary infection, and zoster is due to reactivation of latent VZV acquired during chickenpox. Each disease is characterized by a maculopapular and vesicular skin eruption, which in varicella is generalized, and in zoster is unilateral and usually localized. Varicella is often mild and uncomplicated in otherwise healthy children, but it may unpredictably be associated with significant morbidity and even mortality. In the United States, in the pre-vaccine era, there were about 100 annual deaths from varicella and 11,000 hospitalizations [1]. Most deaths from varicella occurred in individuals who were healthy before contracting varicella. The risk of developing zoster is increased in the immunocompromised and the elderly, and zoster may be severe but is rarely fatal.

Live Attenuated Varicella Vaccine: Background

A live attenuated varicella vaccine, the Oka strain, was developed in Japan in the early 1970s [2]. This vaccine is now available worldwide. The most extensive published experience with varicella vaccine comes from the United States, where it was licensed in 1995 for routine use in healthy susceptible individuals over the age of 1 year [1]. The vaccine is extremely safe [3, 4]. The main adverse effects in healthy persons are both mild and transient: fever and a sore arm in less than 50%, and a very mild rash in about 5%, usually occurring a month after immunization [5]. VZV antibodies are demonstrable after about 2 weeks. In the United States, the epidemiology of varicella is changing, with a decreased incidence of chickenpox in both the vaccinated and the unvaccinated [6].

Transmission of the Oka strain from healthy vaccinees with rash to other susceptibles is extremely rare. It has been reported only on four occasions, although over 30 million doses have been distributed in the United States alone. There is no clinical evidence of reversion of the vaccine strain to virulence. Contact cases are usually very mild or even subclinical [4, 7]. There are many mutations in the vaccine compared to the parental viral strains, although which are responsible for attenuation is unknown [8–10]. Although a few cases of immunodeficiency have been unmasked by complications following varicella vaccination, there have been no reported fatalities from the Oka varicella strain [4].

In the United States, contraindications to varicella vaccine include pregnancy, allergy to vaccine components, and immunodeficiency. Children receiving steroids at a dose of 2 mg/kg/day of prednisone or greater (or of its equivalent) should not be immunized. It is safe to immunize HIV-infected children, as long as their levels of CD4 lymphocytes are relatively well-preserved [11]. Healthy persons who have close contact with susceptible individuals who are at high risk to develop severe varicella, such as pregnant women and children with malignant disease, should be immunized to protect them because transmission of vaccine is rare. The vaccine has been shown to be cost effective in the United States and abroad [12–17].

Efficacy and Effectiveness of the Vaccine

Studies in vaccinated leukemic children showed not only that the vaccine was safe but also that it was highly protective against varicella [18]. About 85% of vaccinated leukemic children were completely protected against varicella after household exposure, and those who developed breakthrough infection had mild infections. Early studies in healthy children indicated a similar degree of protection [19].

Two double-blind placebo controlled studies in healthy children were conducted, indicating protection of about 90% [20–22]. Better protection resulted from higher doses (roughly 3,000 plaque forming units {pfus} vs. 10,000 pfus per dose). A post-licensure case-control effectiveness study in healthy children indicated that the vaccine in the United States is about 85% effective in preventing varicella [23]. The best evidence of protection from varicella vaccine, however, is the reported decline in disease since 1995, in the United States [6].

Does Immunity to Varicella Wane with Time after Immunization?

There are two types of vaccine failure, primary and secondary. Primary vaccine failure means there is no immune response to a vaccine. When a positive immune response after vaccination is lost with time, secondary vaccine failure has occurred. There is little evidence for secondary vaccine failure after immunization of healthy children with the Oka vaccine. Persistence of VZV antibodies and cellular immunity for up to 20 years after vaccination has been reported [24–26]. Studies involving over 400 vaccinated adults for up to 20 years revealed no decrease in immunity to varicella with time. It is disquieting, however, that about 10% of children may develop a modified form of varicella despite vaccination. In some studies, the breakthrough rate of varicella ranged between 18 and 34% [27–30].

Outbreaks of varicella in young vaccinated children have recently been reported in the United States [31–33]. There are a number of possible explanations for these outbreaks. Improper storage of this labile vaccine may account for primary vaccine failure in some children. The ability to mount a protective immune response may be impaired in children with asthma [34]. Children immunized at less than 14 months old may have higher rates of breakthrough varicella than those immunized when they were older [31, 32]. When varicella vaccine is administered less than 1 month after another live vaccine, the incidence of breakthrough varicella increases [35]. A recent outbreak of varicella in a day care center in New Hampshire identified an interval of over 3 years since vaccination as the only significant risk for developing breakthrough disease [33]. This small study is the only one that suggests that waning immunity may be a factor in breakthrough disease. However, the children involved were very young, and the age at vaccination could also have been a factor. Continued investigations are necessary to determine whether waning immunity is significant in developing breakthrough disease.

A second dose of varicella vaccine may prove useful to avoid possible primary and secondary vaccine failure [36]. Breakthrough varicella is almost always mild. The few reports of varicella of normal or increased severity in vaccines probably represent primary vaccine failure [4]. Zoster, however, may be more of a concern. Currently a second dose of varicella vaccine is being considered for all healthy children in the U.S. because of the realization that primary immune failure may be more of a concern than was originally thought [37].

Zoster: Effects and Potential Effects on Its Incidence in the Vaccine Era

In immunocompromised vaccines, the incidence of zoster was lower than after natural infection [25]. Therefore vaccination may also be protective against zoster in healthy children. Less than 50 cases of zoster have been reported after distribution of over 30 million doses of vaccine between 1995 and 2002 in the United States [4].

Recently, it has been proposed that exposure to varicella is protective against zoster, and there is concern that zoster may become more common as the incidence of varicella decreases in a population. Studies in vaccinated leukemic children with either household exposure and/or additional doses of varicella vaccine correlated with a lower incidence of zoster than one dose of vaccine [38]. A recent case-control study indicated that following natural varicella, there is a lower incidence of zoster in individuals who have exposures to children with VZV infections in comparison to those who do not [39]. It has been projected that an epidemic of zoster with accompanying significant mortality will occur in countries where varicella vaccination is routine [40]. These observations, even though theoretical, have led to reluctance to use varicella vaccine routinely in some countries. Since the possibility has been raised, it will be important to study the situation further.

It is important, however, to put the idea of an increase of zoster into perspective. The reported incidence of zoster in healthy individuals aged 40–50 in developed countries is 2–4 cases per 1,000 person-years of observation. It has been projected that the rate of zoster may double in countries with routine vaccination [40]. This could lead to an incidence of 4–8 cases per 1,000 person-years of observation in this age group. This incidence of zoster is still not very high and is about that seen in vaccinated leukemic children. In similar unvaccinated children, the incidence is at least 3 times higher. In adults with AIDS it is 6 times higher. Thus the projected increase in the incidence of zoster based on computer modeling might represent a significant increase, but it is likely to be far from an epidemic. Moreover, the mortality of zoster is lower than that from the primary infection (varicella). As yet, no actual increase in the incidence of zoster has been observed in the United States.

References

1 Centers for Disease Control: Prevention of varicella: recommendations of the advisory committee on immunization practices (ACIP). MMWR Morb Mortal Wkly Rep 1996;45:1–36.

2 Takahashi M, Otsuka T, Okuno Y, Asano Y, Yazaki T: Live vaccine used to prevent the spread of varicella in children in hospital. Lancet 1974;2:1288–1290.

3 Baba K, Yabuuchi H, Takahashi M, Gershon AA, Ogra PL: Seroepidemiologic behavior of varicella zoster virus infection in a semiclosed community after introduction of VZV vaccine. J Pediatr 1984;105:712–716.

4 Sharrar RG, LaRussa P, Galea SA, Steinberg SP, Sweet AR, Keatley RM, Wells ME, Stephenson WP, Gershon AA: The postmarketing safety profile of varicella vaccine. Vaccine 2000;19:916–923.

5 White CJ. Varicella-zoster virus vaccine. Clin Infect Dis 1997;24:753–763.

6 Seward JF, Watson BM, Peterson CL, Mascola L, Pelosi JW, Zhang JX, Maupin TJ, Goldman GS, Tabony LJ, Brodovicz KG, Jumaan AO, Wharton M: Varicella disease after introduction of varicella vaccine in the United States, 1995–2000. JAMA 2000;287:606–611.
7 Tsolia M, Gershon AA, Steinberg SP, Gelb L: Live attenuated varicella vaccine: evidence that the virus is attenuated and the importance of skin lesions in transmission of varicella-zoster virus. J Pediatr 1990;116:184–189.
8 Gomi Y, Imagawa T, Takahashi M, Yamanishi K: Oka varicella vaccine is distinguishable from its parental virus in DNA sequence of open reading frame 62 and its transactivation activity. J Med Virol 2000;61:497–503.
9 Loparev VN, McCaustland K, Holloway BP, Krause PR, Takayama M, Schmid DS: Rapid genotyping of varicella-zoster virus vaccine and wild type strains with fluorophore-labeled hybridization probes. J Clin Micro 2000;38:4315–4319.
10 Gomi Y, Imagawa T: Comparison of DNA sequence and transactivation activity of open reading frame 62 of Oka varicella vaccine and its parental viruses. Arch Virol 2001;S17:49–56.
11 Gershon A, Takahashi M, et al: Live attenuated varicella vaccine; in Plotkin S, Orenstein W (eds): Vaccines, ed 4. Philadelphia, WB Saunders, 2002.
12 Lieu T, Cochi S, Black SB, Halloran ME, Shinefield HR, Holmes SJ, Wharton M, Washington AE: Cost-effectiveness of a routine varicella vaccination program for U.S. children. JAMA 1994;271: 375–381.
13 Beutels P, Clara R, Tormans G, Van Doorslaer E, Van Damme P: Costs and benefits of routine varicella vaccination in German children. J Infect Dis 1996;174:S335–S341.
14 Burnham BR, Wells TS, Riddle JR: A cost-benefit analysis of a routine varicella vaccination program for United States Air Force Academy cadets. Mil Med 1998;163:631–634.
15 Coudeville L, Paree F, Lebrun T, Sailly J: The value of varicella vaccination in healthy children: cost-benefit analysis of the situation in France. Vaccine 1999;17:142–151.
16 Diez Domingo J, Ridao M, Latour J, Ballester A, Morant A: A cost benefit analysis of routine varicella vaccination in Spain. Vaccine 1999;17:1306–1311.
17 Brisson M, Edmunds WJ: The cost-effectiveness of varicella vaccination in Canada. Vaccine 2002;20:1113–1125.
18 Gershon AA, Steinberg SP, Gelb L, Galasso G, Borkowsky W, LaRussa P, Farrara A: Live attenuated varicella vaccine: efficacy for children with leukemia in remission. JAMA 1984;252: 355–362.
19 White CJ: Clinical trials of varicella vaccine in healthy children. Infect Dis Clin North Am 1996;10:595–608.
20 Weibel RB, Neff BJ, Kuter BJ, Guess HA, Rothenberger CA, Fitzgerald AJ, Connor KA, McLean AA, Hilleman MR, Buynak EB: Live attenuated varicella virus vaccine: efficacy trial in healthy children. N Engl J Med 1984;310:1409–1415.
21 Weibel R, Kuter BJ, Guess HA, Rothenberger CA, Fitzgerald AJ, Connor KA, McLean AA, Hilleman MR, Buynak EB: Live Oka/Merck varicella vaccine in healthy children: further clinical and laboratory assessment. JAMA 1985;245:2435–2439.
22 Varis T, Vesikari T: Efficacy of high titer live attenuated varicella vaccine in healthy young children. J Infect Dis 1996;174:330–334.
23 Vazquez M, LaRussa PS, Vazquez M, LaRussa PS, Gershon AA, Steinberg SP, Freudigman K, Shapiro ED: The effectiveness of the varicella vaccine in clinical practice. N Engl J Med 2001;344:955–960.
24 Asano Y, Suga S, Yoshikawa T, Kobayashi I, Yazaki T, Shibata M, Tsuzuki K, Ito S: Experience and reason: twenty year follow up of protective immunity of the Oka live varicella vaccine. Pediatrics 1994;94:524–526.
25 Arvin A, Gershon AA: Live attenuated varicella vaccine. Annu Rev Microbiol 1996;50:59–100.
26 Ampofo K, Saiman L, LaRussa P, Steinberg S, Annunziato P, Gershon A: Persistence of immunity to live attenuated varicella vaccine in healthy adults. Clin Infect Dis 2002;34:774–779.
27 Clements DA, Armstrong CB, Ursano AM, Moggio MM, Walter EB, Wilfert CM: Over five-year follow-up of Oka/Merck varicella vaccine recipients in 465 infants and adolescents. J Pediatr Infect Dis 1995;14:874–879.

28 Johnson, Stancin CT, Fattlar D, Rome LP, Kumar ML: A long-term prospective study of varicella vaccine in healthy children. Pediatrics 1997;100:761–766.
29 Takayama N, Minamitani M, Takayama M: High incidence of breakthrough varicella observed in healthy Japanese children immunized with live varicella vaccine (Oka strain). Acta Paediatr Jpn 1997;39:663–668.
30 Clements D, Moreira SP, Coplan PM, Bland CL, Walter EB: Postlicensure study of varicella vaccine effectiveness in a day-care setting. J Pediatr Infect Dis 1999;18:1047–1050.
31 Dworkin MS, Jennings CE: An outbreak of varicella among children attending preschool and elementary school in Illinois. Clin Infect Dis 2002;35:102–104.
32 Galil K, Fair E, et al: Younger age at vaccination may increase risk of varicella vaccine failure. J Infect Dis 2002;186:102–105.
33 Galil K, Lee B, et al: Outbreak of varicella at a day-care center despite vaccination. N Eng J Med 2002;347:1909–1915.
34 Shapiro E, LaRussa P: Vaccination for varicella – just do it. JAMA 1997;278:1529–1530.
35 Centers for Disease Control: Simultaneous administration of varicella vaccine and other recommended childhood vaccines – United States, 1995–1999. MMWR Morb Mortal Wkly Rep 2001;50:1058–1061.
36 Gershon A: Varicella vaccine: are two doses better than one? N Engl J Med 2002;347:1962–1963.
37 Gershon AA: Control of varicella: why is a two-dose schedule necessary? Pediatr Infect Dis J 2006;25:475–476.
38 Gershon A, LaRussa P, et al: The protective effect of immunologic boosting against zoster: an analysis in leukemic children who were vaccinated against chickenpox. J Infect Dis 1996;173:450–453.
39 Thomas S, Wheeler J, et al: Contacts with varicella or with children and protection against herpes zoster in adults: a case-control study. Lancet 2002;360:678–682.
40 Brisson M, Gay N, et al: Exposure to varicella boosts immunity to herpes-zoster: implications for mass vaccination against chickenpox. Vaccine 2002;20:2500–2507.

Anne A. Gershon, MD
Columbia University College of Physicians and Surgeons
650 W. 168th Street
New York, NY 10032 (USA)
Tel. +1 212 305 9445, Fax +1 212 342 5218, E-Mail aag1@columbia.edu

Gross G, Doerr HW (eds): Herpes Zoster.
Monogr Virol. Basel, Karger, 2006, vol 26, pp 170–188

Appendix: Herpes Zoster Guideline[1] of the German Dermatology Society (DDG)[2]

G. Gross[a], *H. Schöfer*[b], *S. Wassilew*[c], *K. Friese*[d], *A. Timm*[e], *R. Guthoff*[c], *H.W. Pau*[f], *J.P. Malin*[g], *P. Wutzler*[h], *H.W. Doerr*[i]

[a]Universitätsklinik and Poliklinik für Dermatologie und Venerologie, Rostock; [b]Hautklinik, Klinikum der Johann-Wolfgang-Goethe-Universität, Frankfurt a. M.; [c]Städtische Hautklinik, Krefeld, Krefeld; [d]Universitäts-Frauenklinik und Poliklinik; [e]Universitäts-Augenklinik, Universität Rostock, and [f]Universitätsklinik für Hals-Nasen-Ohren-Krankheiten, Universität Rostock, Rostock; [g]Neurologische Klinik, Universität Bochum, Bochum; [h]Institut für Antivirale Chemotherapie, Universität Jena, Jena; [i]Institut für Medizinische Virologie, Klinikum der Johann-Wolfgang-Goethe-Universität, Frankfurt a. M., Germany.

1. Definition

Herpes zoster (zoster, shingles) is a neurocutaneous disease caused by the varicella zoster virus (VZV) as a consequence of declining VZV-specific cell-mediated immunity occurring physiologically with the aging process or by immune suppression and conditions with immunodeficiency. The main risk factor for the development of zoster is increasing age. While the number of cases

[1]Established by an Expert Group on the occasion of a symposium on 'Zoster and Zoster pain' held on 9th December 2000 in Rostock (Germany) consisting of H. Schöfer, Frankfurt a. M.; S. Wassilew, Krefeld (Dermatology and Venereology); K. Friese (Gynaecology and Obstetrics); H.W. Pau (Otorhinolaryngology); A. Timm, R. Guthoff (Ophthalmology); A. Wree (Anatomy), Rostock; J.P. Malin (Neurology), Bochum; H.W. Doerr, Frankfurt a. M. (Virology); P. Wutzler, Jena (Virology); under the leadership of G. Gross, Rostock (Dermatology and Venereology).

[2]Reprinted from Journal of Clinical Virology, V26(3): 277–289, Gross et al.: Herpes zoster guideline of the German Dermatology Society (DDG) © 2003 Elsevier B.V., with permission from Elsevier.

occurring per year is roughly the same over 50 as under 50 years of age, the age-specific incidence of herpes zoster is higher beyond the age of 50 (Hope-Simpson, 1975; Ragozzino et al., 1982; Meister et al., 1998a).

Similar to chronic diseases such as diabetes mellitus, coronary heart disease and depressions, zoster may considerably impair quality of life due to a number of cutaneous complications and complications especially of the nervous system with often chronic sequelae. Zoster is characterised by pain and especially the difficult-to-treat postherpetic neuralgia (PHN).

Growing life expectancy and the increasing number of elderly in Germany has resulted in a higher population risk for zoster and chronic zoster pain. Also, the severe course of the disease in immunosuppressed individuals and organ transplant recipients, tumor patients and AIDS patients, requires effective treatment.

Early antiviral therapy with systemic virostatic agents may shorten healing time of zosteric skin lesions and may prevent in many cases severe pain and other complications.

2. Etiology

Zoster results from reactivation of VZV infection persisting endogenously in the nervous system (Weller, 1983, 1992).

VZV is regarded as the causative agent of both chickenpox and zoster and belongs, like herpes simplex viruses (HSV) type 1 and type 2, to the neurodermatropic herpes viruses. Both VZV and HSV type 1 and type 2 belong to the subfamily alpha herpes virinae of the herpes virus group. Electronmicroscopically these viruses can be hardly differentiated and they also have a number of similarities in molecular biology.

After primary VZV infection leading to chickenpox nucleic acids of VZV remain dormant in sensory ganglia. This phase of latency usually lasts for several decades before VZV reactivation occurs. Unlike HSV, which also remain latent after primary infection, the VZV infection normally presents only with one clinically manifest recurrence, usually in the form of zoster from the middle period of life. Reactivation and viral replication lead to necrosis and inflammation in the affected sensory ganglia. VZVs travel along the sensory nerves to the skin and cause the characteristic painful dermatomal zoster rash with grouped papulovesicular, later pustular skin lesions. Before this usually painful rash develops, prodromal pain and paresthesias will appear in the dermatome affected by the VZV reactivation.

In some cases, inflammation of the affected ganglia causes, via the liquor, a meningitis, mostly without associated encephalitis. After disappearance of both skin symptoms and acute pain, the patients may develop chronic zoster pain which is also described as PHN. In very rare cases, there are no visible cutaneous efflorescences despite acute and chronic zoster pain (zoster sine herpete) (Rudra, 1990).

3. Epidemiology

The reactivation of VZV infection occurs in approximately 20% of seropositive individuals (Balfour, 1988). Normally, the VZV infection recurs only once in life. Immunodeficient patients may develop zoster twice in the same dermatome and, in very rare cases, even several times zoster may recur. In contrast to HSV infection with up to 300 symptomatic and asymptomatic reactivations during lifetime, in which the recurrence risk decreases with age, the risk to develop zoster, increases with age (Meier and Strauss, 1992; Doerr and Rabenau, 1996).

The two world-wide largest studies have described different figures on the incidence of zoster in the general population (Hope-Simpson, 1965; Ragozzino et al., 1982). In an examination of patients from general practices in Great Britain, Hope-Simpson found an incidence of 3.39 per 1000 persons per year, whereas Ragozzino et al. reported 1.3 cases of zoster in 1000 examined persons from Rochester (USA) per year. Principally, zoster may occur in any period of life. The incidence rises steadily until adulthood and then remains constant with 2–3 cases per 1000 per year until the end of the fourth decade of life. Beyond the age of 50, the incidence strongly increases to approximately five cases per 1000 per year. Persons at ages of 60–70 have an incidence rate of 6–7 cases per 1000. Individuals beyond the age of 80 have an incidence of more than 10 cases per 1000 per year. Given these values, more than half of all people who reach 85 years of age will develop zoster at any point of their life (Hope-Simpson, 1965).

Every form of immunodeficiency strongly increases the risk of developing zoster. Fifteen percent of leukemic children (Feldman et al., 1973), 30% of bone marrow transplant recipients (Locksley et al., 1985) and 20% of HIV infected persons within 12 years after seroconversion will experience a VZV reactivation (Veenstra et al., 1995).

HIV infected individuals and adults suffering from cancer have a many times higher zoster incidence than immunocompetent patients of the same age (Friedmann-Kiein et al., 1986; Rustoven et al., 1998; Kost and Strauss, 1996). Leukemic children have a 50–100 times higher incidence than healthy children of the same age (Guess et al., 1985; Balfour, 1988; Glynn et al., 1990; Kost and Strauss, 1996).

4. Cutaneous Manifestation of Zoster

In 80% of patients affected by zoster, the skin manifestation is preceded by a prodromal stage lasting approximately 3–5 days. The symptoms in this phase are not uniform. Patients often complain of tiredness, weariness, mild temperature and other general symptoms. Furthermore, burning pain and paresthesias as well as circumscribed pain in the affected dermatomes are often misinterpreted as myocardial infarction, cholecystis, biliary colic, renal colic, appendicitis and disk prolapse. A condition, in which, after the prodromal stage, dermatome-dependent pain appears without zoster efflorescences, is called zoster sine herpete (Rudra, 1990). The characteristic zoster rash usually affects a single dermatome (zoster segmentalis). However, the dermatomes may overlap each other (Gross, 1997). Crossing of the body's midline is a rarity (so-called zoster duplex). In very rare cases, several skin segments are affected asymmetrically, i.e. on both sides of the body (Vu et al., 1999; Bloss et al., 2001).

In the affected nerve segment, discrete spots develop in stages into a painful, asymmetric, unilateral erythema. Within 12–24 h usually grouped vesicles appear. After another 2–4 days, these vesicles become confluent. On the third day, the vesicles may cloud and then usually dry out over approximately 7–12 days. In immunologically healthy patients, duration of the rash until disappearance of the crusts is usually 2–3 weeks. Occasionally, immunodeficient patients have chronic courses with skin changes lasting for months and repeated vesicular eruptions may develop Hoppenjans et al., 1990; Kost and Strauss, 1996 (Rustoven et al., 1998).

5. Localization of Zoster

Zoster may occur in any dermatome, but most frequent are zoster thoracicus and zoster in the area of the head. Comparison studies showed that the thoracic dermatomes are affected in approximately 50–56% (Hope-Simpson, 1965; Ragozzino et al., 1982; Meister et al., 1998a). The head area, such as the innervation areas of the trigeminal nerve and other cranial nerves (VII and VIII) is involved in approximately 20%. Less frequently affected (in descending order) are cervical, lumbal and sacral segments (Meister et al., 1998a, b). Viral spreading in the blood results in disseminated zoster which is observed in only approximately 1–2% of immunocompetent patients but more frequently in immunodeficient patients (Meier and Strauss, 1992; Cohen et al., 1998). This condition is characterised by nodules and vesicles in a non-dermatomal pattern. These skin lesions may hardly be differentiable from chickenpox. Inner organs such as the lung and especially the nervous system may also be involved.

Zoster is mainly differentiated from other circumscribed rashes by its asymmetry, the synchronous development of skin lesions from erythemas to vesicular, pustular and finally crustous lesions. Varicella-like forms have especially been reported in AIDS and tumor patients. The lesions often do not develop synchronously in stages, thus creating a problem for differential diagnosis (Galagher and Merigan, 1979; Hoppenjans et al., 1990).

6. Symptoms

In most cases, the clinical appearance of zoster is preceded by prodromal symptoms such as pain or paresthesias. Pain that occurs before, during or after the dermatomal rash stage is called zoster-associated pain. Postzosteric or PHN is defined as pain which appears or persists after the cutaneous symptoms (Dworkin et al., 1997).

Different mechanisms in the pathogenesis of chronic zoster pain are being discussed: inflammatory nerve lesions with destruction of peripheral nerve structures or neurones in the sensory ganglia as well as modified signal processes in the central nervous system which occur during and subsequent to VZV reactivation (Kost and Strauss, 1996; Malin, 1996).

7. Complications and Sequelae of Zoster

Acute and chronic complications affecting skin, eye, and central nervous system are rather frequent (Table 1), whereas complications in inner organs are relatively rare.

In the acute stage, the skin is primarily affected by bacterial secondary infections that can be accompanied by ecthymiform ulcerations. Other major cutaneous complications are haemorrhagias (zoster haemorrhagicus), purulent gangrene (zoster gangraenosus), persistence of lesions and dissemination (zoster disseminatus). The latter occur especially in immunodeficient patients. Chronic inconvenient sequelae on the skin are hypo- and depigmented-scars, more seldom granulomatous reactions and manifestation of a psoriasis vulgaris (Köbner's phenomenon) (Gross, 1997).

Ophthalmologic complications comprise zoster lesions of the eyelid with subsequent scar formation, conjunctivitis, episcleritis, scleritis, corneal involvement with epithelial and stromal keratitis as well as endothelitis. Furthermore, uveitis may occur with the risk of secondary glaucoma or even acute retinal necrosis. The optic nerve can also be affected (Liesegang, 1991).

Table 1. Complications of zoster

	Cutaneous and mucocutaneous sites	Nervous system	Eye	Inner organs
Acute complications	Bacterial secondary infections Zoster haemorrhagicus Zoster gangraenosus Zoster generalisatus	Encephalitis Meningitis Granulomatous arteritis Segmental pareses Facial paresis in zoster oticus	Conjunctivitis Episcleritis/scleritis Uveitis Keratitis Iridocyclitis (→glaucoma)	Pneumonia Esophagitis Myocarditis Enterocolitis Pancreatitis
Chronic complications	Persisting zoster Scar formation (atrophic scars, hypertrophic scars) Hypo/depigmentation Granulomatous skin lesions Pseudolymphoma Manifestation of psoriasis (Köbner's phenomenon)	PHN Guillain-Barré syndrome Myelitis Motor neuropathy Abdominal hernias Phrenoplegia Bladderdysfunction	Keratitis Chorioretinitis Retrobulbar neuritis Vasculitis Panophthalmitis Atrophy of optic nerve	

Source: Gross (1997)

Chorioretinits and neuritis of the optic nerve are more frequently observed in AIDS patients. Also, herpetic retinal necroses due to VZV infection with the risk of bilateral blindness were reported in HIV positive patients.

Other neurologic complications comprise zoster meningitis, motor neuropathies and paralysis, Guillain-Barré syndrome, granulomatous arteritis and affection of cranial nerves (Malin, 1996). The most frequent and important complication of zoster affecting the nervous system, is acute and chronic pain also known as PHN. PHN is defined as pain that persists for longer than 4 weeks or that occurs 4 weeks after a pain-free interval. Approximately 10–20% of zoster patients of all ages are affected by PHN. Chronic zoster pain in children is extremely rare (Malin, 1996).

The manifestation rate of PHN shows a clear age-dependence. In zoster patients over 55, 60 and 70 years it is 27, 47 and 73%, respectively. In women and patients with zoster ophthalmicus, PHN seems to occur more frequently. Immunodeficiency seems to be no risk for the development of chronic pain (Hope-Simpson, 1975; Guess et al., 1985; Gross, 1997).

Best documented complications are postzosteric neuropathies and paralysis such as phrenoplegias, abdominal hernias, bladder dysfunction and cystitis. In some cases these symptoms may be misinterpreted (Malin, 1996). Another neurologic complication is segmental paralysis which has a good outcome, since in every second case complete regression occurs.

8. Zoster in Immunodeficient Patients

Zoster may present atypically in patients with pathologic or iatrogenic immunodeficiency. The rash may occur mitigated with few symptoms. However, in most cases, distinct inflammation, partly with hemorrhages and occasionally with necroses, is found. In some cases, more than one dermatome is involved and the rash may persist for a longer period than in immunocompetent patients.

Disseminated zoster with varicella-like skin involvement and involvement of inner organs is observed more frequently in immunodeficient patients (Cohen et al., 1998; Rustoven et al., 1998). Zoster is regarded as an early marker for HIV infection (Melbye et al., 1978; Friedmann-Kiein et al., 1986; Schöfer, 1991). Therefore, HIV infection should always be excluded serologically in zoster patients younger than 50.

With reduced cellular immunity, the zoster rash may be associated with atypic varicelliform or even verrucous or ecthymiform lesions (Galagher and Merigan, 1979; Hoppenjans et al., 1990; Schöfer, 1991; Schöfer et al., 1998).

9. Diagnosis

Zoster is usually diagnosed by inspection with an asymmetrical dermatomal rash and grouped vesicles. Further evidence is found with the rash limited by the midline, its dermatomal arrangement and, especially, prodromal and zoster-associated segmental pain. Diagnostic problems may occur particularly in the prodromal or early phase of zoster when only erythematous skin lesions exist.

9.1. Laboratory Diagnosis

The differentiation of the VZV infection from herpes simplex and bullous dermatoses is an important indication for virological diagnosis. Also, VZV infections of pregnant women and of newborn infants, atypical infections of immunodeficient patients and suspected VZV infection of the central nervous system must be confirmed by laboratory diagnosis.

Today, the VZV polymerase chain reaction (PCR) as well as the direct detection of virus in cell cultures and the detection of specific antibodies to VZV in special circumstances are recommended.

The enzyme-linked immunosorbent assay and the immunofluorescence technique are especially suited for the detection of VZV-specific immunoglobulins of classes IgG, IgM and IgA. VZV–IgG rises may occur spontaneously and in recurrent HSV-infections due to cross reactivity of epitopes. However, additional detection of IgM and high-titered IgA anti-VZV antibodies usually indicates a reactivated VZV-infection regardless whether lesions are visible or not. Generally the patient consults a physician 1–3 days after the occurrence of symptoms. Thus clinical diagnosis is mostly supported by serologic findings (Doerr et al., 1987; Wutzler and Doerr, 1995; Doerr and Rabenau, 1996). Serological diagnosis is particularly efficient and helpful in immunodeficient patients suffering from VZV reactivation (Wutzler and Doerr, 1995).

In very early zoster (erythematous stage), detection of VZV infection in tissue by the VZV-specific PCR may occasionally be helpful (Lilie and Wassilew, 1999). Additionally, VZV infection can be detected using the immunofluorescence test. Another indirect but less sensitive detection procedure is the so-called Tzanck test that examines the cytopathic effect of VZV in the epidermis with characteristics multinucleate giant cells and intranuclear inclusions (Barr et al., 1977; Solomon et al., 1986; Gross and Doerr, 1997). Unfortunately, this cytopathic effect is not VZV-specific since it is also seen in cutaneous HSV-infections.

Electron microscopy allows the morphological detection of herpes viruses in vesicular fluids or smears. However, this again does not allow differentiation of the herpes viruses VZV, HSV-1 and HSV-2 (Barr et al., 1977; Gross and Doerr, 1997). Furthermore electron microscopy cannot be used routinely.

9.2. Differential Diagnosis

The major differential diagnoses of zoster are the zosteriform herpes simplex and the different forms of erysipelas, such as hemorrhagic and bullous erysipelos.

Further differential diagnoses are contact dermatitis, insect bites, bullous dermatoses such as the bullous pemphigoid and pemphigus vulgaris. Phlegmones and panniculitis pose less frequently differential diagnosis problems. In cases where the clinical presentation is atypical, particularly if no symptoms exist, viral detection should be done by PCR, or viral culture as well as by serology and also by the Tzanck test (Gross and Doerr, 1997; Lilie and Wassilew, 1999).

If chronic pain persists after the zoster lesions have healed diagnosis can be confirmed retrospectively by serologic tests. Viral detection is not possible in this phase.

10. Therapy

The aim of zoster treatment is to relieve pain in the acute phase, to limit the spread and duration of zosteric skin lesions and to prevent or alleviate PHN and other acute and chronic complications (Table 1). It is of utmost importance that ophthalmological complications should be referred for specialist ophthalmic management as soon as possible. Complications related to zoster of other cranial nerves such as zoster oticus always require consulting a specialist.

10.1. Symptomatic Therapy of Zoster

Depending on the stage of the rash, treatment will be done locally either through drying and antisepsis, e.g. with wet dressings, with lotio alba, vioform zinc mixture or later by crust removal. A satisfactory local therapy with proven antiviral efficacy does not exist. Local zoster therapy with antiviral substances has shown to be ineffective and is not recommended (Gross, 1997).

Painlessness should be achieved by appropriately-dosed analgesics (e.g. tramadol), often in combination with a neuroactive agent, (e.g. amitriptylin) (Malin, 1996). However, it is unknown as to whether systemic analgesia prevents the development of PHN in the early stage of zoster.

Table 2. Indications for systemic antiviral therapy of zoster

Urgent indications
Zoster of any localization in patients beyond the age of 50
Zoster in the head/neck area of patients at any age
Severe zoster on the trunk/on the extremities
Zoster in immunodeficient patients
Zoster in patients with severe atopic dermatitis and severe eczemas
Relative indications
Zoster on the trunk/on the extremities in patients younger than 50 years

10.2. Indications for Antiviral Chemotherapy

Zoster is a self-limiting disease affecting skin and nervous system.

Even without specific antiviral treatment, the circumscribed zoster on the trunk and on the extremities of young individuals without risk factors usually heals without complications.

Antiviral treatment shortens the healing process and is especially important when a complicated clinical course is expected (Table 2).

Systemic antiviral therapy is urgently indicated in patients beyond the age of 50, in immunodeficient patients, in patients with malignant primary disease as well as in patients with involvement of cranial nerves, especially with affection of the first branch of trigeminal nerve (zoster ophthalmicus) but also with zoster oticus. Furthermore, urgent indications for systemic antiviral treatment are severe dermatitis atopica and other severe eczemas. It is noteworthy that zoster ophthalmicus and zoster oticus as well as an age beyond 50 are also major risks for the development of PHN.

Furthermore, antiviral therapy should always be initiated if vesicles develop on more than one skin segment. Patients with hemorrhagic lesions and/or patients with mucosal affection should also always receive systemic antiviral treatment (Table 2). The Zoster Study Group of the Dermatological Research Cooperative Group (ADF) of the German Dermatology Society and of the Paul-Ehrlich-Society developed a zoster score (Meister et al., 1998a, b) which can be used as decision aid for the initiation of systemic antiviral therapy. This score takes into account the following risk factors.

Age beyond 50, dermatomal pain, female gender, more than 50 efflorescences, hemorrhagic efflorescences, involvement of cranial or sacral dermatomes. The practicability of this zoster score system in clinic routine has been demonstrated at the Department of Dermatology and Venereology of the University of Rostock (Schlecht et al., in preparation).

10.3. When to Start Systemic Antiviral Therapy

The success of each antiviral treatment is dependent on the starting point of therapy. Systemic virostatic therapy must be initiated as soon as possible, i.e. if possible within 48 to a maximum of 72 h after onset of skin symptoms. The effect level of the antiviral drug must be achieved rapidly and maintained for an optimum therapeutical success. In the following situations, systemic antiviral therapy can also be started even at a later date (after 72 h).

Disseminated zoster with evidence of immuno-deficiency and affection of inner organs, persisting zoster ophthalmicus and zoster oticus. Furthermore, even after 72 h, antiviral treatment of zoster is urgently indicated in all patients with known immunodeficiency. Some more recent data suggest that valacyclovir, the prodrug of acyclovir, continues to be beneficial in preventing PHN even if given 72 h after onset of symptoms.

10.4. Antiviral Therapy of Zoster

A total of 4 different systemic antiviral substances for treatment of zoster have become available in Germany: acyclovir, valacyclovir, famciclovir and brivudin, which very recently has been approved (illustration), whereas in the United States only three drugs – acyclovir, valacyclovir and famciclovir – are in use (Gnann and Whitley, 2002).

All of these substances can be given orally (Gross and Laskowski, 2001). Only acyclovir can be administered both enterally and parenterally. Up to now, valacyclovir and famciclovir have been the world-wide most commonly used antiviral drugs for the oral treatment of uncomplicated zoster. Better pharmacokinetics, better bioavailability and easier application make them superior to oral acyclovir. A recent double-blind randomised study showed valacyclovir and famciclovir to have similar effects on zoster-associated pain and on PHN in immunocompetent patients (Tyring et al., 2000). Valacyclovir and famciclovir have to be administered orally 3 times daily for 7 days, whereas oral acyclovir must be given in a dose of 800 mg, 5 times daily. Brivudin has the great advantage of requiring only once daily dosing. In VZV infections, brivudin has a markedly higher antiviral potency than acyclovir, valacyclovir and famciclovir (De Clercq et al., 1979). Brivudin, when given once daily (1 × 125 mg for 7 days), is therefore well effective and superior to oral acyclovir treatment (5 times daily dosing) (Table 3).

This is especially true for patients beyond the age of 50 who are at a higher risk of developing PHN. Patients treated with brivudin had significantly faster stop of viral replication compared to patients treated by oral

Table 3. Current antiviral therapy for zoster

Valacyclovir oral	1000 mg	3 × daily	7 days
Acyclovir oral	800 mg	5 × daily	7 days
Acyclovir intravenous	5–7.5 mg	3 × daily	7 days
Acyclovir intravenous[a]	8–10 mg	3 × daily	7–10 days
Famciclovir oral	250 mg	3 × daily	7 days
Brivudin oral	125 mg	1 × daily	7 days

[a]Zoster in immunodeficient patients.

acyclovir (5 × 800 mg per day) (Wutzler et al., 2001). Already in 1995, Wutzler et al. were able to show that oral administration of brivudin, 125 mg 4 times daily, is as effective as intravenous acyclovir, 10 mg per kg body weight with regard to acute zoster symptoms in immunodeficient patients (Wutzler et al., 1995).

Recently, a randomised observation study showed an 11% lower PHN rate in the patient group treated with brivudin (32.7%) than in the acyclovir group (43.5%) ($P = 0.006$). As with all observation studies this study carried certain drawbacks and the observation needs further evaluation (Wassilew et al., 2001a). A restriction applies to immunosuppressed patients, children and pregnant or lactating women who cannot be treated with brivudin. Furthermore, brivudin must not be used in combination with 5-fluorouracil or with other drugs containing 5-fluoropyrimidines. In the last year, brivudin has been approved in Germany for oral treatment of zoster, due to the markedly more convenient administration, brivudin (125 mg once daily) seems to be superior, especially in the elderly, to the oral antiviral drugs valacyclovir (1000 mg, 3 times daily) and famciclovir (250 mg, 3 times daily).

A large multicenter, prospective, double-blind, randomised study in 2027 patients showed brivudin (125 mg, once daily) to be at least as effective as the antiviral famciclovir (250 mg, 3 times daily) for prevention of PHN. Both PHN duration as well as zoster-associated pain were influenced by once daily brivudin in the same way as with the triple dosed famciclovir (Wassilew et al., 2001b). Currently recommended antiviral therapy in Germany is shown in Table 3.

10.5. Intravenous Antiviral Therapy

Parenteral therapy with acyclovir (5–10 mg per kg body weight, 3 times per day) is the standard treatment for zoster in immunodeficient patients.

However, in contrast to studies comparing intravenous acyclovir versus oral Brivudin (Wutzler et al., 1995), no controlled studies comparing intravenous acyclovir with oral treatment of acyclovir, valacyclovir or famciclovir have been performed for this patient group so far.

The doses of intravenous acyclovir, oral acyclovir, oral valacyclovir and oral famciclovir have to be adapted for patients with reduced renal function. In contrast, even with reduced creatinine clearance, it is not necessary to change the dosage of oral brivudin.

10.6. Treatment of Zoster Affecting Cranial Nerves

There are no controlled studies comparing intravenous acyclovir in zoster ophthalmicus and zoster oticus with oral antiviral treatments such as acyclovir, valacyclovir, brivudin or famciclovir. The maintenance of a sufficient virostatic plasma level is of utmost importance especially for the treatment of zoster of these locations. Therefore, in most cases, intravenous acyclovir therapy under stationary conditions has to be preferred to oral antiviral therapy.

10.6.1. Ophthalmological Complications

Systemic intravenous or oral antiviral treatment of zoster ophthalmicus must be done as soon as possible. For oral treatment, acyclovir 800 mg 5 times a day for at least 7 days and valacyclovir 1000 mg 3 times a day, also for 7 days are recommended. Colin et al. showed that the latter valaclovir dose has the same efficacy as acyclovir (5 times 800 mg per day) in the prophylaxis of ocular complications of zoster such as conjunctivitis, keratitis and pain (Colin et al., 2000). Longer treatment may occasionally be needed. However, controlled studies showed no significantly better clinical efficacy when acyclovir was given for 14 or 21 days (Hoang-Xuan et al., 1992; Beutner et al., 1995).

Viral epithelial keratitis must be treated with local virostatic agents (e.g. acyclovir eye ointment). It is recommended to refrain completely from steroid therapy, except in endothelitis and trabeculitis. In this case, systemic combination therapy consisting of acyclovir and prednisolone is recommended. The steroids dose should be carefully considered to balance between the antiviral effect and the tissue-damaging immune reaction (Sundmacher, 1996).

10.6.2. Zoster Oticus

Zoster oticus results from an infection of ganglial cells of the VII and VIII cranial nerves. Clinically, it is characterised in most cases by severe earache, loss of hearing (conductive deafness), vertigo and/or facial paresis. After ENT consultation and specific otological diagnosis, the following therapy is usually indicated: high dose antiviral therapy (intravenously preferred) in combination with glucocorticoids, infusion treatment with rheologics or physiological saline solution with vasoactive substances. Analgesics and, in cases with strong vertigo, anti-vertigo agents are additionally indicated.

10.7. Corticosteroid Therapy

The additive zoster therapy with high dose steroids shortens the duration of acute zoster pain, but has no essential effect on chronic pain (PHN). This is the result of 2 large prospective studies (Wood et al., 1994; Whitley et al., 1996). The use of corticosteroids must be carefully considered, especially in view of possible side effects.

It is not recommended to use corticosteroids without systemic antiviral therapy. Study results are available only for a combination therapy of prednisolone (40 and 60 mg per day, dose reduction over ≈ 10–14 days to 5 mg per day) and acyclovir. However, nothing speaks against possible combination of prednisolone with valacyclovir, brivudin or famciclovir.

10.8. Therapy of Neuralgia: Step-by-Step Scheme

Patient-specific aspects and side effects are important considerations in pain therapy. Early presentation to a pain therapist or a pain outpatient clinic may be required.

1) *Step 1*: non-steroidal analgesics (e.g. paracetamol 1.5–5 g per day)
2) *Step 2*: additional low-potency opioid analgesics (e.g. tramadol 200–400 mg per day, codeine 120 mg per day), if necessary, combined preparations.
3) *Step 3*: in addition to a 'peripheral' analgesic, administration of a high-potency central opioid (e.g. buprenorphine 1.5–1.6 mg per day; oral morphine 30–360 mg per day) is indicated. This refers to patients who fail to respond the more measured treatment approaches.

In severe neuralgic pain, step 1 or 2 in combination with an anti-convulsant (carbamazepine 400–1200 mg per day). Anti-convulsants (carbamazepine) can reduce the lancinating pain, but are not effective for continuous pain. Another option is gabapentine (900–2400 mg per day). Also, anti-depressants, amitryptilin 10–75 mg) and neuroleptics (levomepromazin 20–150 mg per day) may be efficient, especially in elderly patients (Malin, 1996; Wassilew, 2000).

Further treatment possibilities consist in local therapy with capsaicin, local anesthetic blocking of sympathetic nerve, transcutaneous electric nerve stimulation, if necessary, neurosurgery (e.g. thermocoagulation of substantial gelatinosa Rolandi) in exceptional cases.

10.9. Therapy of Other Neurologic Complications of Zoster

Zoster meningitis, zoster encephalitis and zoster myelitis should be treated with parenteral acyclovir in a dose of 10 mg per kg body weight, 3 times daily. Also in severe zoster ophthalmicus and zoster oticus, intravenous acyclovir should be preferred to enteral therapy.

10.10. Therapy of Zoster in Children and Adolescents

Zoster in childhood is usually not a strict indication for systemic antiviral therapy. An exception is zoster affecting cranial nerves. Also, children with hereditary or acquired immunodeficiency as well as atopic children might benefit from antiviral therapy. As long as there are no controlled treatment studies in children with immunodeficiency and zoster available, one can only speculate whether antivirals, given early, might prevent viral spread, reduce the severity of inflammation and prevent secondary immunosuppression. It is likely, but not yet proven, that severe tissue destruction, secondary bacterial infections and scar formation, might be reduced by an early antiviral therapy. In analogy of the sometimes severe, but individually not predictable course of zoster in patients with HIV infection we suggest to treat immunodeficient and atopic children with acyclovir intravenously.

10.11. Therapy of Zoster in Immunodeficient Patients

Only acyclovir has been approved so far for the treatment of zoster in severely immunocompromized patients. Other oral virostatic agents such as brivudin, valacyclovir and famciclovir showed positive results in case reports especially in HIV-infected patients, but are not officially approved. In a given case, the responsible physician may use these drugs after careful consideration of the risks involved.

It has been shown that the therapeutical regimen of acyclovir is dependent on the severity of immunodeficiency and clinical findings. If the CD4 cells are roughly within limits of normal (>400 cells per μl), segmental zoster may be treated with a standard dose of acyclovir intravenous (5–7.5 mg per kg body weight every 8 h). In severe immunodeficiency with extended skin lesions, especially when neurologic symptoms are present, treatment with high dose intravenous acyclovir (10 mg per kg every 8 h) and continuous monitoring of renal function is required. In severe immunodeficiency with extended skin lesions, especially when neurologic symptoms are present, treatment with high dose intravenous acyclovir (10 mg per kg every 8 h) is required. As acyclovir has some nephrotoxic properties and cumulates in patients with impaired renal function, serum creatinine clearance must be tested with the first given dose of acyclovir. In case of reduced serum creatinine clearance the interval to the following acyclovir infusions must be prolonged from 8 to 12 or even 24 h (according to the nomograms or tables provided by the producers of acyclovir).

Acyclovir resistance has to be assumed if the clinical findings improve only slowly or not at all. In such cases, another treatment option would be intravenous foscarnet which, however, is highly nephrotoxic (Breton et al., 1998).

Apart from the constant number of HIV patients in Germany (≈ 1500–1700 new infections in 2000), especially the prophylaxis and therapy of other immunodeficient patients such as organ transplant recipients, patients under immunosuppressive agents and tumor patients strongly require therapeutics that can be used if acyclovir resistance has developed. Also, zoster patients with renal damage (contraindications for acyclovir and foscarnet) require new non-nephrotoxic virostatic agents.

10.12. Therapy of Zoster in Pregnancy

Unlike varicella, maternal zoster is hardly considered a risk for the unborn. Normally, there is no viremia that could pose a risk to the development of the unborn. Antiviral therapy of the mother with intravenous acyclovir (3 times 5–10 mg per kg body weight for 7 days) is indicated only in exceptional cases. If possible, zoster during pregnancy should be treated locally and symptomatically. Topical acyclovir is definitely not recommended. The treatment of a pregnant zoster patient in a gynaecologic practice should be in such a way that non-pregnant women without immune protection are not put at risk of varicella.

11. Resistance of Varicella Zoster Virus to Antivirals

VZV can become resistant to nucleoside analogues such as acyclovir, valacyclovir, famciclovir and brivudin, probably due to mutations in the thymidin kinase (TK) gene or mutations in the polymerase gene. Viral strains with mutations in the TK gene are usually resistant to acyclovir, famciclovir and also to ganciclovir. Such virus strains have been isolated from several HIV infected patients who had been suffering from chronic VZV infections and who had been treated with acyclovir for longer periods. Acyclovir-resistant VZV strains have no TK or no proper TK due to their altered substrate specificity. Usually, this mechanism cannot be by-passed by increasing the dose. In such cases, treatment of choice is intravenous foscarnet, 3 times 40 mg or 2 × 50 mg per kg body weight per day (Breton et al., 1998). However, foscarnet may also be ineffective in polymerase gene mutations. The only alternative would then be treatment with intravenous cidofovir (Safrin et al., 1997).

Appendix A: Illustration

Acyclovir	Valacyclovir
Penciclovir	Famciclovir
Brivudin	Foscarnet

Systemic antiviral substances for zoster therapy approved in Germany.

References

Balfour HH. Varicella zoster virus infections in immunocompromised hosts. Am J Med 1988;85:68–73.

Barr RJ, Hertens J, Graham JH. Rapid method for Tzanck preparation. J Am Med Ass 1977;237: 1119–20.

Beutner KR, Friedman DJ, Forszpaniak C, et al. Valacyclovir compared with acyclovir for improved therapy for herpes zoster in immunocompetent adults. Antimicrob Agents Chemother 1995;39: 1546–53.

Bloss G, Ebisch MA, Kunz M, Gross G. Bilateraler asymmetrischer Zoster im Jugendalter. Hautarzt 2001;4:335–8.

Breton G, Fillet AM, Katlama C, et al. Acyclovir-resistant herpes zoster in human immunodeficiency virus-infected patients: results of foscarnet therapy. Clin Infect Dis 1998;27:1525–7.

Cohen RR, Beltraini VP, Grossman ME. Disseminated herpes zoster in patients with human immunodeficiency virus infection. Am J Med 1998;84:1076–80.

Colin J, Prisant O, Cochener B, et al. Comparison of the efficacy and safety of valacyclovir and acyclovir for the treatment of herpes zoster ophthalmicus. Ophthalmology 2000;107:1507–11.

De Clereq E, Descamps J, De Somer P, et al. (*E*)-5-(2 Bromovinyl)-2′-deoxyuridine: a potent and reactive anti-herpes agent. Proc Natl Acad Sci USA 1979;76:2947–51.

Doerr HW, Rabenau H. Dermatotrope Herpesviren, Infektionsbiologie, Epidemiologie und Diagnostik. Chemotherapie J 1996;1:1–8.

Doerr HW, Rentschler M, Schleifler G. Serologic detection of active infections with human herpes viruses (CMV, EBV, HSV, VZV): diagnostic potential of IgA class and IgG subclass-specific antibodies. Infection 1987;15:93–8.

Dworkin RH, Carrington D, Cunningham A, et al. Assessment of pain in herpes zoster: lessons learned from antiviral trials. Antivir Res 1997;33:73–85.

Feldman S, Hughes WT, Kim HY. Herpes zoster in children with cancer. Am J Dis Child 1973;126: 178–84.

Friedmann-Kiein AE, Lafleur FL, Gendler E, et al. Herpes zoster: a possible early sign for development of acquired immunodeficiency syndrome in high-risk individuals. J Am Acad Dermatol 1986;14: 1023–8.

Galagher JG, Merigan TC. Prolonged herpes zoster infection associated with immunosuppressive therapy. Am Intern Med 1979;91:842–6.

Glynn C, Crockford G, Garaghan D, et al. Epidemiology of shingles. J R Soc Med 1990;83:617–9.

Gnann JW, Whitley RJ. Herpes zoster. N Engl J Med 2002;347:340–6.

Gross G. Zoster. Dtsch med Wschr 1997;122:132–9.

Gross G, Doerr HW. Labordiagnose dermatotroper Virusinfektionen: Erregerisolierung. In: Korting HC, Sterry W, editors. Diagnostische Verfahren in der Dermatologie. Berlin, Wien: Blackwell Wissenschaftsverlag; 1997, p. 197–202.

Gross G, Laskowski J, Virustatika. In: Korting HC, Sterry W, editors. Therapeutische Verfahren in der Dermatologie, Berlin, Wien: Blackwell Wissenschaftsverlag; 2001, p. 647–54.

Guess HA, Broughton DD, Melton LJ, Kurland LJ, III. Epidemiology of herpes zoster in children and adolescents: a population-based study. Pediatrics 1985;76:512–7.

Hoang-Xuan T, Büchi R, Herbort CP, et al. Oral acyclovir for herpes zoster ophthalmicus. Ophthalmology 1992;99:1062–70.

Hope-Simpson RE. The nature of herpes zoster: a long term study and new hypothesis. Proc R Soc Med 1965;58:9–20.

Hope-Simpson RE. Postherpetic neuralgia. J R Coll Gen Practit 1975;25:571–5.

Hoppnjans WB, Bibler MR, Orne RL, et al. Prolonged cutaneous herpes zoster in acquired immunodeficiency syndrome. Arch Dermatol 1990;126:1048–50.

Kost RG, Strauss SE. Postherpetic neuralgia – pathogenesis, treatment and prevention. N Engl J Med 1996;335:23–42.

Liesegang TJ. Diagnosis and therapy of herpes zoster ophthalmicus. Ophthalmology 1991;98:1216–29.

Lilie HM, Wassilew SW. Shingles (Zoster). In: Wolff MH, Schünemann S, Schmidt A, editors, Varicella-zoster-virus. Molecular biology pathogenesis and clinical aspects, vol. 3. Contrib Microbiol Basel Karger, 1999;111–27.

Locksley RM, Flournoy N, Sullivan KM, et al. Infection with varicella-zoster virus after marrow transplantation. J Infect Dis 1985;152:1172–81.

Malin JP. Zoster und Nervensystem. Dtsch med Wschr 1996;121:635–8.

Meier JL, Strauss SE. Comparative biology of latent varicella-zoster virus and herpes simplex virus infections. J Inf Dis 1992;166(Suppl):S13–23.

Meister W, Neiss A, Gross G, et al. Demography, symptomatology and course of disease in ambulatory zoster patients. Intervirology 1998a;41:272–7.

Meister W, Neiss A, Gross G, et al. A prognostic score for postherpetic neuralgia in ambulatory patients. Infection 1998b;26(6):359–63.

Melbye M, Grossman RJ, Goebert JJ, et al. Risk of AIDS after herpes zoster. Lancet 1978;1:728–30.

Ragozzino MW, Melton LJ, Kurland TL, et al. Population based study of herpes zoster and its sequelae. Medicine (Baltimore) 1982;61:310–6.

Rudra T. Zoster sine herpete. Brit J Clin Pract 1990;44:284.

Rustoven JJ, Ahlgren P, Elhakim T, et al. Varizella zoster infection in adult cancer patients: a population study. Arch Intern Med 1998;148:1561–6.

Safrin S. Cherrington J, Jaffe HS. Clinical uses of cidofovir. Rev Med Virol 1997;7:145–56.

Solomon AR, Rasmussen JE, Weiss JS. A comparison of Tzanck smear and virus isolation in varizella and herpes zoster. Arch Dermatol 1986;122:282–5.

Sundmacher R. Therapie der Viruserkrankungen des vorderen Augenabschnittes. In: Kampik A, editors. Das äußere Auge: Hauptreferate der XXXI—Essener Fortbildung für Augenärzte, Bücherei des Augenarztes; 1996, Vol. 137, p. 121–5.

Schlecht K, Laskowski J, Gross G. Zoster-Scoresystem zur Risikoabschätzung bezüglich der Entwicklung einer post-zosterischen Neuralgie-Erste Anwendungen in der Praxis. In preparation.

Schöfer H. Frühsymptome der HIV-Erkrankung an Haut- und Schleimhäuten. AIDS-Forschung (AIFO) 1991;6:633–48.

Schöfer H, Baur S, Gregel C, et al. Hyperkeratotic varicella zoster virus infection in a HIV-infected patient. Successful treatment of persistent lesions with cryosurgery. Br J Dermatol 1998;138:714–5.

Tyring SK, Beutner KR, Tucker BA, et al. Antiviral therapy for herpes zoster. Randomized, controlled clinical trial of valacyclovir, and famciclovir therapy in immunocompetent patients of 50 years and older. Arch Fam Med 2000;9:863–9.

Veenstra J, Krool A, van Praag RM, et al. Herpes zoster, immunological deterioration and disease progression in HIV-1 infection. AIDS 1995;9:1153–8.
Vu AQ, Radonich MA, Heald PW. Herpes zoster in seven disparate dermatomes (zoster multiplex): report of a case and review of the literature. J Am Acad Dermatol 1999;40:868–9.
Wassilew SW. Differentialtherapie der Herpesvirusinfektionen. In: Plettenberg A, Meigel WN, Moll I, editors. Dermatologie an der Schwelle zum neuen Jahrtausend: Springer-Verlag; 2000, p. 315–8.
Wassilew SW, Schumacher K, Städtler G, et al. A randomised double-blind survey on the effect of brivudin in the prevention of postherpetic pain in comparison with acyclovir. Fourth International Conference on Varicella, Herpes Zoster, Postherpetic Neuralgia. La Jolla, USA; 2001, March, p. 3–5.
Wassilew SW, Schumacher K, Macher K, Städtler G, et al. Brivudin compared to famciclovir in the prevention of postherpetic neuralgia: a randomised double-blind multi-center trial. Fourth International Conference on Varicella, Herpes Zoster, Postherpetic Neuralgia. La Jolla, USA; 2001, March, p. 3–5.
Weller TH. Varicella and herpes zoster: changing concepts of the natural history, control and importance of a not-so-benign virus. N Engl Med 1983;309:1362–8.
Weller TH. Varcella and herpes zoster. A perspective and overview. J Infect Dis 1992;166(Suppl 1):1–6.
Whitley RJ, Weiss H, Gnann IW, et al. Acyclovir with and without prednisone for the treatment of herpes zoster. A randomised placebo-controlled trial. The National Institute of Allergy and Infectious Diseases Collaborative Antiviral Study Group. Ann Intern Med 1996;125:376–83.
Wood MJ, Johnson RW, Mc Kendrick MW, et al. A randomised trial of acyclovir for 7 days or 21 days with and without prednisolon for treatment of acute herpes zoster. N Engl J Med 1994;330:896–900.
Wutzler P, Doerr HW. Der Herpes zoster-ein Morbus herpes non simplex. Dtsch med Wschr 1995;120: 1133–8.
Wutzler P, de Clercq E, Wutke K, Färber I. Oral brivudin vs. intravenous acyclovir in the treatment of herpes zoster in immunocompromised patients: a randomised double-blind trial. J Med Virol 1995;46:252–7.
Wutzler P. Stubinski BM, Koch I, et al. Brivudin compared to acyclovir in the treatment of acute herpes zoster: a randomised, double-blind, multicenter trial. Fourth International Conference on Varicella, Herpes Zoster, Postherpetic Neuralgia. La Jolla, USA; 2001, March, p. 3–5.

Gerd Gross, MD
Professor of Dermatology and Venereology, Director
Department of Dermatology and Venereology
University of Rostock
Augustenstrasse 80–84
DE–18055 Rostock (Germany)
Tel. +49 381 4949701, Fax +49 381 4949702, E-Mail gerd.gross@med.uni-rostock.de

Subject Index